DATE DUE

AG 4 '94			

REPRODUCTIVE HAZARDS IN THE WORKPLACE: MENDING JOBS, MANAGING PREGNANCIES

Regina Kenen, PhD, MPH

SOME ADVANCE REVIEWS

"A very comprehensive, insightful, and accessible book on an issue of great urgency—the impact of workplace conditions on the health and rights of women and future generations. Kenen provides basic tools (questions to ask, resources to turn to) so that women and men can better make their way through complex occupational terrains. Especially helpful are concrete portrayals of reproductive hazards associated with various jobs and profiles of what can be done to abate those hazards and to protect worker rights."

Lin Nelson, PhD, Environmental Studies, Evergreen State College; Co-Chair, Committee on Occupational Health of the National Women's Health Network

"Documents and explains the wide variety of workplace hazards faced by twentieth-century workers in both blue and white collar industries. Kenen's observations on the political implications of the new fetal protection policies are incisive. Readers will be delighted with her attention to solutions as well as to the hazards themselves. A must for unions, employer groups, and the insurance industry, not to mention working people everywhere."

Kate Short, Total Environment Centre, Sydney, Australia

"Gain an understanding of the potential for reproductive hazards in the working environment affecting the woman, man, and child. . . . This book debunks myths and prejudices about working during pregnancy. It espouses the basic and explicit principle of a pregnant woman's right to work in a way that does not compromise her health or that of her unborn child. . . . Aims to motivate people to become . . . more active in their efforts to create a more healthful workplace for all.

Chloë Mason, PhD, Consultant on Occupational Health & Safety, Discrimination Issues, and Environmental Protection, Sydney South, Australia

---

NOTES FOR PROFESSIONAL LIBRARIANS AND LIBRARY USERS

This is an original book title published by The Haworth Press, Inc. Unless otherwise noted in specific chapters with attribution, materials in this book have not been previously published elsewhere in any format or language.

CONSERVATION AND PRESERVATION NOTES

The paper used in this publication meets the minimum requirements of American National Standard for Information Sciences—Permanence of Paper for Printed Material, ANSI Z39.48-1984.

Reproductive Hazards in the Workplace

Mending Jobs, Managing Pregnancies

HAWORTH Women's Studies
Ellen Cole, PhD and Esther Rothblum, PhD
Senior Co-Editors

New, Recent, and Forthcoming Titles:

When Husbands Come Out of the Closet by Jean Schaar Gochros

Prisoners of Ritual: An Odyssey into Female Genital Circumcision in Africa by Hanny Lightfoot-Klein

Foundations for a Feminist Restructuring of the Academic Disciplines edited by Michele Paludi and Gertrude A. Steuernagel

Hippocrates' Handmaidens: Women Married to Physicians by Esther Nitzberg

Waiting: A Diary of Loss and Hope in Pregnancy by Ellen Judith Reich

God's Country: A Case Against Theocracy by Sandy Rapp

Women and Aging: Celebrating Ourselves by Ruth Raymond Thone

Women's Conflicts About Eating and Sexuality: The Relationship Between Food and Sex by Rosalyn M. Meadow and Lillie Weiss

A Woman's Odyssey into Africa: Tracks Across a Life by Hanny Lightfoot-Klein

Anorexia Nervosa and Recovery: A Hunger for Meaning by Karen Way

Reproductive Hazards in the Workplace: Mending Jobs, Managing Pregnancies by Regina Kenen

Women Murdered by the Men They Loved by Constance A. Bean

Our Choices: Women's Personal Decisions About Abortion by Sumi Hoshiko

Reproductive Hazards in the Workplace

Mending Jobs, Managing Pregnancies

Regina Kenen, PhD, MPH

The Haworth Press
New York • London • Norwood (Australia)

The Haworth Press, Inc., 10 Alice Street, Binghamton, NY 13904-1580

Library of Congress Cataloging-in-Publication Data

Kenen, Regina.
Reproductive hazards in the workplace : mending jobs, managing pregnancies / Regina Kenen.
p. cm.
Includes bibliographical references and index.
ISBN 1-56024-154-3 (acid free paper)
1. Reproductive toxicology. 2. Industrial toxicology. I. Title.
RA124.2.K46 1992
618.3052 – dc20

91-34582
CIP

To Joanne, Marc, Stephanie and Judith,
my own ‘‘reproductive hazards,’’
and to my husband Peter who helped create them.

ABOUT THE AUTHOR

Regina Kenen, PhD, MPH, is Professor of Sociology at Trenton State College in Trenton, New Jersey, and Co-chair of the Occupational and Environmental Health Committee, National Women's Health Network, Washington, DC. Dr. Kenen is the author of numerous articles and co-author of the book *Turning Things Around: A Women's Occupational and Environmental Health Guide*. She is the mother of four children.

CONTENTS

PART II. THE WORKPLACE: HOW SAFE IS SAFE?

PART III. STRATEGIES FOR THE FUTURE

Photo © Earl Dotter

Preface

Women throughout the country are more aware than ever before of occupational and environmental hazards. They have heard accurate and inaccurate new reports about possible harmful effects of asbestos insulation crumbling in ceilings, vapors emitted by photocopying machines, anaesthetic gases leaking in operating rooms, hairdyes and cosmetic preparations used in beauty salons, and chemical plant spills. Women no longer are satisfied by many of the answers they receive regarding the safety of the factories, offices, laboratories, hospitals, schools, and stores in which they and their families work.

Furthermore, the number of pregnant women and mothers with children under 18 in the paid work force has burgeoned. According to a federal survey in the 1980s, 59% of all women bearing their first child and 22% of those bearing their second or later child became pregnant while they were employed outside the home. Of these, about 50% stayed on the job until their seventh month of pregnancy or longer. In 1988, 1.7 million employed women gave birth. By the early 1990s, it is expected that 80% of employed women will become pregnant during their careers.

Physical, chemical, and biological substances as well as psychological and social pressures can harm pregnant workers or make their pregnancies more difficult. Workplace hazards may cause miscarriages, stillbirths, and an increase in the number of children born with physical abnormalities, mental retardation and learning disabilities. Furthermore, children of exposed workers may face a greater risk of developing childhood cancer. Some of these reproductive threats, such as exposure to lead, are well documented. Others are strongly suspected. Still others, such as the effects of working long hours at video display terminals, have been reported by women, but are still under investigation. Much is still unknown.

The special importance of reducing occupational risks for the pregnant worker is part of the much larger social objective of protecting all women's occupational health and safety. This book was written to help women attain this goal. It aids working women in understanding and making decisions about pregnancy and job related health problems. It helps them judge scientific data and assess risks, and provides them with the kind of information needed to make their decisions. It gives them a standard for judging their own work situation, shows how they might improve it, and how, armed with increased knowledge, they can improve working conditions for all pregnant women.

While this book alerts women to possible dangers in their work place, *it is not a substitute for obtaining individual health care advice from a well-qualified professional*. Every worker's physiological and genetic makeup differs as does their interaction with the work environment. Several occupational health clinics located in schools of public health or medical centers are listed in the resource organization section in Appendix B.

It is equally important that workers realize that they must be aware of tomorrow's hazards as well as today's. New technologies can present potential reproductive hazards in the workplace, and these need to be continually monitored and evaluated. Working women – and men – need to understand and use scientific information to promote and preserve their own and their unborn children's health. The ability to protect their health during their reproductive years may be as important to them as learning specific job skills.

Women are not alone in their vulnerability to the damaging influence of working conditions on their children. Men are exposed to occupational reproductive hazards as well. More and more evidence suggests that men's reproductive organs and sperm are suffering damage in the workplace and that their offspring are suffering the consequences. The act of conception is a joint enterprise between women and men. Freeing the workplace from reproductive hazards should also be a joint venture. Potential fathers (and employers) will find the information in this book beneficial.

This book draws on information from public health and scientific literature, unions, government documents and public interest groups. But it is the voices of the pregnant women themselves who were interviewed that speak most dramatically of the urgent need to achieve a "pregnancy-friendly" workplace.

Readers have diverse agendas. Each of you may be concerned about different aspects of reproductive hazards in addition to learning about the specific occupational risk you face. You may be distressed about the dearth of scientific information, or you may want to learn about the protection government regulatory agencies offer, or the success of social action. Therefore, several facets of the problems are covered and each chapter is self-contained. You may just want to read the chapters pertaining to your workplace or individual concerns instead of reading the entire book. You may also want to return to other chapters later when you feel the need to investigate further.

Chapter 1 surveys some of the historical hazardous working conditions under which women have labored that affected their reproductive functioning. Most jobs held by women were in the past and still are, low paying and non-unionized. Therefore, women have had little say in improving their lot. Labor legislation designed to protect women by excluding them from night work and heavy lifting was also used to exclude them from any jobs that were no more hazardous than "women's work" but were better paid. This chapter also discusses the difficulty working women still have in achieving the right to a safe reproductive health environment, pregnancy benefits, job security and child care facilities. The United States is far behind the rest of the major industrial countries in these areas.

Chapter 2, reviewing the biology of reproduction, examines evidence showing that males, as well as females, face reproductive risks on the job which can damage their reproductive systems. These hazards also increase the rate of birth defects and cancer among their offspring and the likelihood of their female partners having miscarriages. The conclusion reached is that a policy that only tries to protect women by excluding them from certain jobs with reproductive risks results in exposing men to harm. Information is given about pregnancy hotlines and prenatal diagnostic techniques.

The impact of the physical and social work environment on the pregnant woman is addressed in Chapter 3. For example, a pregnant woman's body undergoes physical changes that make her more sensitive to noise and heat. In the final months of pregnancy, her balance shifts and she may be uncomfortable if she has to work for many hours in one position. The treatment she receives from her supervisors and co-workers also influences her well-being. Minor, relatively inexpensive changes can often eliminate stress and discomfort so that work remains a pleasurable and healthy activity.

Women need to know whether their specific jobs pose hazards to themselves or their unborn children. Chapters 4 and 5 analyze a variety of work sites—home, office, school, hospital, factory, and the service industry. These chapters identify and suggest ways of minimizing or avoiding hazards in these environments. Only after women are armed with knowledge about the reproductive health risks from their occupational exposures can they decide what steps they want to take to avoid them. For this purpose, an occupational information form for you to fill out and give to your doctor or midwife is included in Chapter 4.

The fact that scientists and consumers view risks differently, and that management may manipulate scientific data makes it very difficult for workers to judge risks intelligently. Chapter 6 talks about how to weigh risks and benefits. It focuses on the information gap and how to close it. Obtaining information and being confident of its accuracy is a major problem for the pregnant worker. The media often exaggerate stories while corporations frequently hide information. Even when adequate data are available, the individual may not know how to put information together to obtain a clear picture of the situation. Chapter 6 translates technical vocabulary and scientific procedures into plain English. Once pregnant women familiarize themselves with scientific terms and risk analysis, they can understand the dangers they face at work and can question and counter explanations, denials and claims made by their employers.

Several initiatives taken in the United States and other countries provide useful insights and valuable suggestions about ways of making work safe and healthy. The final chapter recounts some of the innovative approaches taken by women, including successful organizing and lobbying techniques, and presents an ideal model for the reduction of work-related reproductive health hazards instituted in Sweden.

The glossary at the back of the book defines commonly used

terms in the occupational health field. The description of protective legislation and governmental regulatory agencies (Appendix A) and organizational resources (Appendix B), and the bibliography of occupational health references provide up-to-date sources of information. This book can be referred to repeatedly in the journey towards achieving a pregnancy-friendly workplace.

Acknowledgements

I want to thank my colleagues, friends, students, family, and professionals in the fields of occupational and women's health whose suggestions and encouragement made this book possible. My students Vilma Amadio, Bonnie Baker, Chris Castana, Sandy Collins, Barbara Harned, Vidya Mahajan, Holly Mittleman, Ruth Patterson, Tristen Spears, and Barbara Towle who so diligently conducted the interviews deserve a particular vote of thanks.

Veronica Muller critiqued numerous chapters in an effort to help me find the "right voice." Lin Nelson went over an earlier version with a fine-tooth comb and Dan Kass kindly checked Chapters 3, 4 and 5 for accuracy and clarity. I am particularly grateful for their efforts. Catherine Avent, Kris Barlow, Sue Barlow, Claire-Marie Fortin, Susan Klitzman, Chloë Mason, Marilyn Lauglo, Marsha Love, Daphne May, Judy Norsigian, Maureen Paul, Eugenia Shanklin, Sybil Shainwald, and Jeanne Stellman gave me much useful advice on how to improve the book and how to find a publisher. I am delighted that their endeavors were successful. Joanne Kenen, my daughter, who is an author and journalist, exhorted me to write clearly and "not like an academic." I hope her advice bore fruit. My husband Peter critiqued and edited the manuscript and designed the format for the myriad lists of information presented in the book. Probably his hardest job was living with me while I was writing the manuscript.

Last, but not least, I am grateful to the Faculty Institutional Research and Sabbatical Leave Committee which awarded me released time from teaching to work on this book.

PART I.

PREGNANCY AND THE WORKING WOMAN

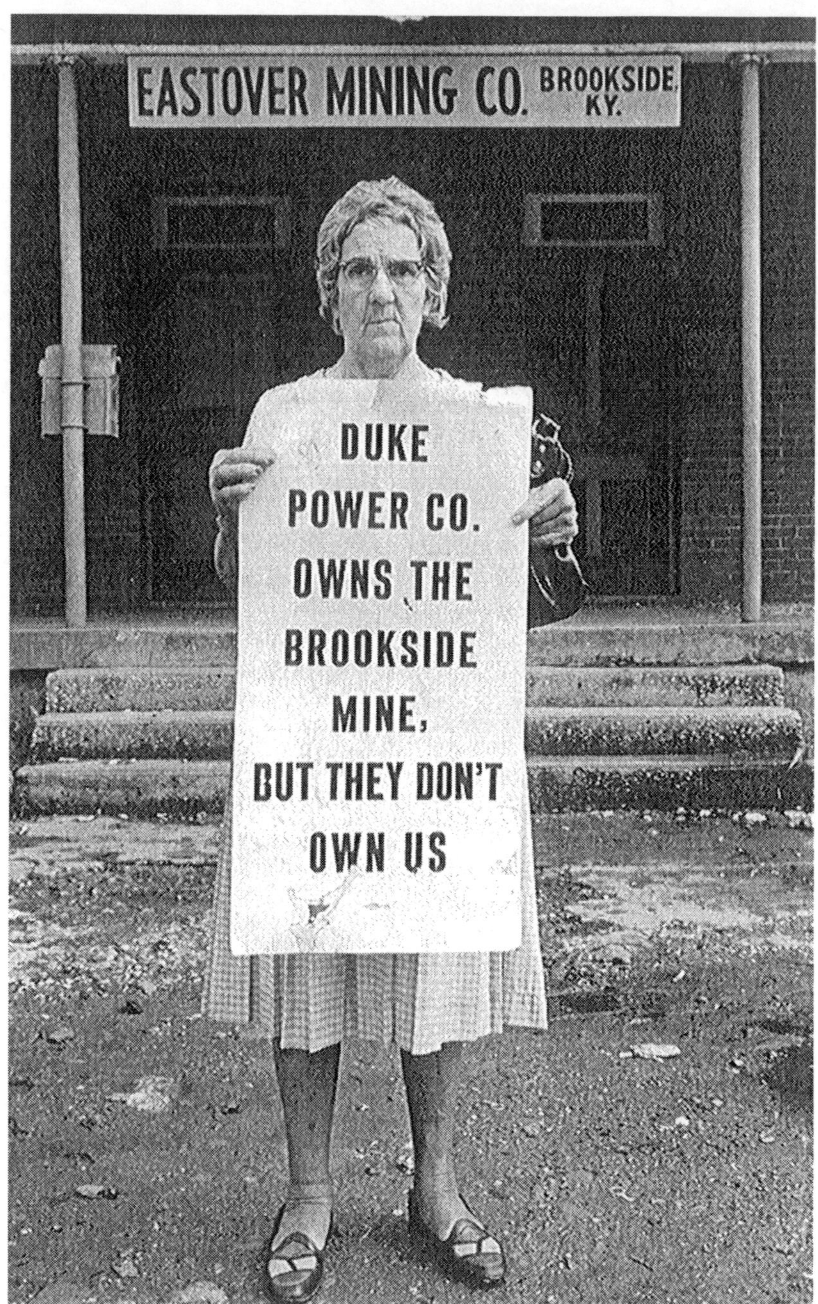

Photo © Earl Dotter

Chapter 1

Working Women: Then and Now

The Factory Girl

(Stanza 4)

Ten hours a day of labor,
  In a close, ill-lighted room.
Machinery's buzz for music,
  Waste gas for sweet perfume;
Hot, stifling vapors in summer,
  Chill drafts on a winter's day,
No pause for rest or pleasure
  On pain of being sent away;
So ran her civilized serfdom –
  FOUR CENTS an hour the pay!

J. A. Phillips in *Machinists' Monthly Journal*
September, 1895, p. 3

Work is as much a part of a woman's life today as is marriage, pregnancy, and motherhood. This is not really new, although men sometimes act as if it were. Except for the rich, women have always worked either in their homes or in outside employment; only the patterns of work have shifted. For example, the now middle-aged mothers of today's young working women arranged their parenting functions and work force participation in sequential order – first working hard in the home and then expanding into the paid employment market. Their daughters, however, are attempting to do both simultaneously. Their work and that of their partners affect their present and future reproductive lives.

Concern about reproductive hazards is nothing new. Almost every pregnant woman has questioned aloud, or asked herself, whether her baby will be all right. What is different is that in the past poor women were powerless to do much about the reproductive hazards associated with work, and the more affluent, middle-class, educated, and vocal women who worked either did not have children, left work when they became pregnant, or believed, sometimes erroneously, that they worked in a safe environment. These social trends have changed and women no longer accept resignedly that some of their offspring will be stillborn, that others will not survive till adulthood, and still others will be forced to cope with a physical or mental handicap.

Couples today expect to have a small number of "quality" children. This is, perhaps, an unrealistic expectation whose fulfillment is too frequently thwarted. The media has given a great deal of attention to the rise of infertility, increases in the number of newborns with birth defects, and heartbreaking stories of attempts by infertile couples to use new reproductive technology in order to have a "child of their own." The causes of these reproductive problems are unknown and widely debated. But possible culprits responsible for the 10-20% of couples having trouble conceiving and the doubling (from 2% to 4%) over the past 25 years of infants born with mental and physical handicaps may be exposures of the mother during pregnancy (and in some instances, the father before conception) to one or several toxic substances or agents. These often are found in the workplace and the environment.

Reproduction in the broadest sense, from ovulation to nurturing, from sperm genetics to economic responsibility for our children, is a social as well as biological phenomenon and may be constrained by work conditions. Not too long ago women were forced to leave their jobs as soon as their pregnancies became evident. These discriminatory employment policies were based on social attitudes rather than on reliable bodies of scientific information. Men were embarrassed by pregnant women's bodies. Now, though the medical profession has given its blessing for healthy pregnant women to work throughout most or all of their pregnancies, and pregnant women are being accepted by men as co-workers, not enough is being done to ensure that their work is safe and comfortable.

OUR ANCESTORS: THE GOOD AND THE BAD

Being able to understand what our mothers, grandmothers, great-grandmothers, and even great-great-grandmothers faced provides us with a sense of continuity with the struggles of the past and teaches us the values of patience and persistence.

The early colonial days were good for white European women settlers. They could own land, run small shops, and trade. The shortage of skilled labor and family system of production encouraged enlightened fathers and husbands to teach their skills to wives and daughters. The idea that the man was the sole supporter of his wife was not common. Instead, husband and wife were considered to be dependent on each other and both responsible for the support of the children. This is very similar to the modern conception of marriage. Instead of the joint family system of production prevalent in Colonial days, we have substituted a joint responsibility for the family system of providing outside income earned in the paid labor force.

This rosy picture, however, was far from universal. If you were a white indentured servant or a black slave, life was much harder. For example, if an indentured servant had a child by a man who would not, or could not, buy out her remaining term, she could be required to serve as much as two additional years to pay for her reduced services and as punishment for her sin. This was true even if she could prove that her own master was father of the child. Slave owners, on the other hand, welcomed rather than condemned the pregnancy of one of their slaves. The birth of a child added to the owner's wealth because he owned his black workers and their children for their lifetimes.

Our Ancestors Didn't Always Have It So Good

Conditions for working women became worse during the industrial revolution in Europe and America. Pregnant workers and those with young children had a particularly difficult time. At first it was single women who took jobs in mills and factories, while married women took in laundry or boarders to make ends meet, or did piece-work at home while they watched their children. Later mothers as well as their daughters entered the outside work force in order to buy food and clothing for their families. Poor women never had the choice between "work" and family. They had to do both, working long hours and often under very unsafe conditions.

> I was a weaver, and we had a lot of lifting to do. My first baby was born before its time, from me lifting my piece of the loom on to my shoulder. . . . Two of us had them to lift, then carry them from the shed across the yard to be weighed. If I had been able to take care of myself I should not have had to suffer as I did for seven weeks before that baby was born and for three months after; and then there was the baby suffering as well, as he was a weak little thing for a long time . . . And when I had my second baby I had to work all through again, as my husband was short of work and ill at the time. I had to go out to work again at the month end, and put the baby out to nurse. I had to get up by four in the morning, and get my baby out of bed, wash and dress it, and leave home by five, as I had half an hour walk to take my baby to my mother's and then go to my work and stand all day till half-past five at night, and then the walk home again with my baby.
>
> Women's Cooperative Guild, *Maternity: Letters from Working Women*, Belmont, London, 1915, pp. 107-108

> The first part of my life I spent in a screw factory from six in the morning till five at night; and afterward used to do my washing and cleaning. I only left two weeks and three weeks before my first children were born. After that I took in lodgers and washing and always worked up till an hour or so before baby was born. The results are that three of my girls suffer with their insides. None are able to have a baby. One dear boy was born ruptured on account of my previous hard work.
>
> . . . I had three little ones in two years and five months, and he (my husband) was out of work two years, and during that time I took in washing and sewing, and have not been near a

> bed for night after night. I was either at my sewing machine or ironing after the little ones had gone to bed. After being confined five days I have had to do all for my little ones. I worked sometimes up till a few moments before they were born.
>
> Women's Cooperative Guild, *Maternity: Letters from Working Women*, Belmont, London, 1915, pp. 161-162

The ongoing workplace misery experienced by working class pregnant women knew no geographic boundary. In Paris, midwives who attended the female tobacco workers often remarked at how frequently the pregnant cigar and cigarette makers aborted. They felt that the only way it was possible for a pregnant tobacco worker to carry her child to term was to temporarily give up her employment. Otherwise, even if carried to term, the infant would either be born dead or would die soon after birth. One French doctor confirmed the midwives' observations. He recorded 45 miscarriages among the cigar and cigarette makers in one district alone.

Another French physician, M. Constantin Paul, published detailed descriptions of the outcome of fifteen pregnancies of four women during their employment in a type foundry. Ten ended in spontaneous abortion, two in premature labor, one in a stillborn, and one in a living child, who died a few hours after birth.

Not only were there reproductive hazards on the job, but women returned too soon to tiring, strenuous and dangerous work, further damaging their reproductive potential. The average period of time off from work after childbirth was five or six weeks; many women among the most highly skilled factory employees in Birmingham generally returned to work after a month.

The toll on the women's mothering capacity and the lack of adequate child care further undermined the health of the workers' children.

> Under all circumstances dust-sorting is dirty and disagreeable work. It is generally undertaken by women of the lowest class. . . . The work is hard and exposed. It not only unfits women for other employment, but even for the ordinary duties of housewife and motherhood. It destroys the best instincts of maternity. The work takes the mother away from her children,

> who are consequently ill tended and often die from neglect. Although several children may be born of these women, in many instances, none of them live beyond a few months, not by the employment affecting the children through the mother, but because maternal duties are totally disregarded. For these reasons, therefore, it is scarcely desirable work for women under thirty years of age. It is with dust women as with many of the laundry women in London; they form a class by themselves and so the work becomes more or less hereditary.
>
> Thomas Oliver, ed. *Dangerous Trades*, John Murray, London, 1902, p. 280

The culture of poverty repeated itself at the end of the 19th century just as it does today and in the same locale—the inner city slums.

In his 1902 treatise, Dr. Oliver cites cases of this maternal deprivation found among children of poor women working in the mills and factories as well. He describes the irony of the situation. The mothers worked so that they could afford more food for their families. But his evidence indicates that the health of the children would have improved if their mothers had stayed home to care for them even if that meant having less to eat. Too often the inadequate caretakers were aged overworked women or "older" brothers or sisters who themselves were only five or six years old. The deaths and diseases befalling these neglected children can be included as reproductive hazards of the workplace as can the plight of their work-worn and care-worn mothers.

> Young, single women also faced reproductive injury caused by working conditions that jeopardized their general health before pregnancy. This occurred not only in mills and factories but in so-called lady-like occupations, as was the case in the abuse of the shop girls in Victorian and Edwardian England.
>
> The "living in" system was developed ostensibly to protect the morals of the girls. But like many other supposedly protective efforts, they more often exploited than protected.
>
> In the spring of 1883 thirty young "ladies" . . . were obliged to sleep four in a bed in a room where there was no

door, the room being on a level with the young men's sleeping apartment, also having no door.

Surtherst, Thomas, *Death and Disease Behind the Counter*, 1884, p. 225

My sleeping-room was on the ground floor, with the window facing on to the street, so that one could step into the street from the window and back into the room from the street, without going through the door at all. I was put into a room with a woman of mature age who had a life of a most undesirable kind; that was my first experience of a living-in house.

Report of the Truck Committee, 1909, vol. 4, III, p. 19

So much for the vaunted protection of purity and virtue. While many of these businesses treated their shop girls decently, a substantial minority were callous and inhuman. They worked the girls from 8 a.m. until 9 or 10 p.m., with no half-day off. They then locked them out of the house on Sunday without providing them with any food. If a girl became ill or pregnant, she was thrown out without any Christian charity.

Some can be traced going from church to church, to find one which is warmed, where they can rest and doze through the time of service before again setting out on their weary pilgrimage. In all weathers these girls may be seen in the cemeteries on Sundays, glad of a place to sit down in. . . . In the places to which they resort to buy food, the acquaintances which they must make, and the snares into which they fall, are only the natural result of a state of things which is a disgrace to any Christian country.

"Philanthropic Labour in the Metropolis," *Wesleyan Methodist Magazine*, Feb., 1873, pp. 182-183

Life was so harsh and the future so bleak in Europe that between 1880 and 1920 several million people immigrated to the United States seeking a better life. But conditions were onerous for women in America too. Dreams may have survived, but the grim reality for the new immigrant women was just trying to make ends meet by living in squalid, over-crowded tenements and working in hazardous jobs.

Take the case of an average immigrant family in 1907. The family could survive on about $11.50 a week, but the father as an unskilled worker could earn only $10. Other members of the family, in order to make up this $1.50 deficit, frequently had no other option than to work in poorly lit, poorly ventilated, crowded and unsafe factories. Complaining about these conditions did no good as the bosses either ignored the complaints or fired the complainers.

THE TOLL OF REPRODUCTIVE HARM

While little was done to ameliorate reproductive hazards in the work environment, awareness of the risks began to slowly emerge. A few voices like Dr. Thomas Oliver and Dr. Alice Hamilton provided evidence pointing to the severe reproductive effects of many industrial substances. There was also a gradual accumulation of knowledge about reproductive harm that was passed on from generation to generation and is now known to have been accurate.

For example, ever since Roman days, it was believed that smoke from extinguished candles could bring about premature labor. This smoke contained carbon monoxide. Dr. Hamilton determined that amounts of carbon monoxide in the mother's blood too slight to produce obvious symptoms in her own body might be enough to kill her fetus. Dr. Hamilton believed that such poisoning might be very common in those industries employing many women. Other investigations showed that lead caused stillbirths, miscarriages, and sterility, and that lead, arsenic, and mercury contaminated the mother's milk. Even the occupation of the father was suspected as playing a role. As far back as 1902, Dr. Thomas Oliver wrote that even though lead had a stronger effect on the reproductive health of women it was also capable of diminishing the virility of men. He reported that babies whose mothers and fathers were both lead workers would be so puny and ill-nourished that they would either be born dead or die a few hours after birth. (The effects of lead exposure on male workers are still being minimized today, particularly by those employed by industries using lead processes.)

Another dangerous substance was benzene, an ingredient in cement, that many women in factories used in gluing parts together. At that time it was not generally known that pregnant women are much more susceptible to benzene poisoning when they are pregnant than when they are not. Dr. Hamilton tells of a woman who worked with benzene for seven years without any noticeable problems but almost lost her life when she was pregnant. She began to have headaches, dizziness, nausea, and pain, so she quit work in her eighth month. She delivered normally, but afterwards returned to the hospital with a high fever, rapid pulse, severe anemia, and kidney disease. She alternated between being in a stupor and delirium. Her recovery took over seven months.

The death rates for mothers during childbirth and for infants were unreasonably high. In some mill towns, the rates were 75% higher for infants born to mothers working in the mills than for those born to women who did not work at all, and 50% higher than for those who worked at home.

Horror stories fueled the drive towards protective legislation. Attention was mainly centered around the long hours women worked and the conditions under which they worked. Much less concern was shown about either the chemical hazards they faced or the particular susceptibility of pregnant workers.

PROTECTIVE LEGISLATION

Two types of laws were proposed. One kind of proposal was aimed at regulating working conditions and the second kind was aimed at excluding women from certain kinds of jobs and from working during times of the night considered to be dangerous for them. Working women differed in their attitudes about how beneficial or how harmful some of these proposals really were.

The first set of laws aimed at providing women with standards for safe and clean working conditions, minimizing health hazards, shortening hours, providing a minimum wage, and compensating them for job-related accidents. These laws were what the legislators felt women would have sought for themselves if they had the bargaining power, and were laws that would protect male workers as well. The second type of protective labor laws was restrictive and aimed almost exclusively at women. For example, laws preventing women from working at night, supposedly for their own protection, actually excluded them from some skilled and well-paying jobs in the printing trade and in railway employment. Despite the appalling conditions the shop girls worked in during the Victorian era, women shopwork reformers feared legislation aimed at protecting women only. They advocated only slow and quiet legislative changes for fear that any limitations such as those restricting the number of hours that women could work would put them at a disadvantage compared with men. Currently, fetal protection policies instituted by many companies (discussed later in this chapter) fulfill the same purpose.

Near the end of the nineteenth century and the beginning of the twentieth century, when the public became more aware of the inhuman working conditions in mines, mills, and other industries, reformers' pressure forced governments to assess the health costs of the industrial revolution to their workers, particularly women and children. A few European countries passed laws to protect the vulnerable pregnant worker and new mother as well.

In 1889 Belgium passed a law stating that women must not be employed in industry within four weeks after child-birth. In 1891 the Factory Act in Britain also stated that new mothers would not be allowed to return to work until one month after delivery, and in 1901 Denmark passed legislation that women could not be employed within four weeks of childbirth except on the presentation of a medical certificate stating that the mother's employment would not injure herself or her child.

In 1877 Switzerland passed a regulation forbidding pregnant women from working in a factory for eight weeks prior to childbirth and eight weeks after. In 1897 this law was amended to forbid employment of pregnant women in certain dangerous occupations for even longer periods of time prior to delivery. These included processes that produced fumes of white phosphorus, or that used lead or lead products, mercury, or sulfuric acid; employment in dry cleaning and India rubber works; and work involving lifting or carrying heavy weights, or the risk of violent shocks. Of course, no mention was made of provisions to pay the woman for the time she was forbidden to work. Presumably, she would not have worked under such adverse conditions until time of delivery or returned to work prematurely unless she needed her wages desperately.

In 1900 Spain passed legislation that prohibited the employment of women within three weeks of childbirth. In addition, Spain was the only country to legislate that mothers be allowed time to nurse their infants during their working hours and be paid for the nursing time as well. Employers had to allow these nursing mothers one hour during the ordinary period of employment to nurse their infants (for which the employers could not deduct wages). The mother was allowed to divide this hour into two separate absences of half an hour. She could choose the time of day that she needed to nurse and her only obligation was to notify her supervisor as to when that was. This legislation was ahead of its time. Now nearly 90 years after the Spanish and Swiss laws were enacted, the United States still has made very little significant progress in designing an equitable and safe policy for pregnant and nursing workers.

THE LINGERING INFLUENCE OF THE WAR YEARS

Not until World War I did women workers gain widespread recognition. Initially during the First World War and then again during the Second World War, women proved how successfully they could handle men's traditional jobs even in the heavy industries that had been closed to them during peacetime. It was the treatment women received during these war years that set the stage for the demands made by employed women today. Companies like Kaiser shipbuilding wooed young mothers into their work force by providing child care, on-site shopping and banking facilities, hot lunches and cooked take-home dinners. Despite the labor shortage, however, pregnant women were considered unfit for work. The usual practice was not to hire pregnant women and to discharge those already employed when their pregnancies became known. Thus many working women hid their pregnancies as long as possible. By doing so they may have unintentionally exposed their fetuses to hazardous substances, particularly during the early vulnerable months.

Issues that the war had dramatized—equal pay, child care, and community support for wage-earning women—receded from the public eye after the war, and women again faced poorly paid jobs or no jobs at all. Moreover, when married women did take advantage of the few opportunities open to them, this often created family stress as the question of whether or not married women should work was once again hotly debated. Pregnant employees fared even worse because working during pregnancy once you "showed" was considered an embarrassment to supervisors, co-workers and customers. Equality for women workers was the foremost demand at the time, but it was not until many years later that anti-discrimination standards were to be applied to pregnant workers as well.

The unfairness of unequal access to jobs and unequal pay for equal work became more galling. Women had attempted to obtain legislative relief several times to no avail. They tried again in the early 1960s and this time they succeeded. In 1962 the Kennedy administration outlawed discrimination in the federal civil service, and in 1963 it pushed through Congress an Equal Pay Act prohibiting different pay for men and women working at equivalent jobs.

The Civil Rights Act of 1964 was the turning point in the quest for equal opportunity and equal treatment in the work force and is the mainstay of women's legal employment rights today. Ironically, the original bill did not even include protection of women. Sometimes momentous historical decisions depend on quirks of fate. Virginia's senator Howard Smith threw the word "sex" into Title VII of the act as a joke to show how ridiculous it was for the federal government to issue hiring practice criteria to private employers. When passage seemed likely, the committee tried to remove the word. Luckily, two politically savvy and tough-minded women legislators came to the rescue. Senator Margaret Chase Smith and Representative Martha Griffiths threatened to stall the bill unless the word "sex" was left in. Thus "sex" became part of a clause that prohibited firms with fifteen or more employees from discriminating on the basis of religion, race and ethnicity. (Senator Howard Smith might have tried even harder to get the word "sex" out, if he could have foreseen that the sex protection clause would in the future be expanded to include pregnant workers as well.)

In 1964 the Equal Employment Opportunity Commission was also created to administer Title VII. The Commission was surprised by the immediate flood of complaints about sexual discrimination and responded to them by formulating guidelines that employers could follow to avoid sex discrimination charges.

Another landmark year for women's rights was 1978. This time pregnant women were the victors. Congress passed the Pregnancy Disability Amendment to Title VII of the 1964 Civil Rights Bill. Pregnancy was considered to be a disability and thus covered under a company's disability insurance, as would be any other injury or illness that prevented an employee from working. While the Amendment is inadequate because it does not cover the vast majority of pregnant working women whose employers do not carry disability insurance, the willingness to make a start on protecting pregnant workers' jobs was a significant advance. (A brief summary and discussion of protective legislation and governmental regulatory agencies can be found in Appendix A).

Considering a healthy pregnancy as a disability may seem a bit odd, but political considerations were at issue. Many of the women pushing for pregnancy benefits legislation cited the long history of discrimination against hiring women. They were afraid to draw attention to the special needs of the pregnant worker. Instead they emphasized the similarity between men and women and compared pregnancy with any medical condition that temporarily disabled a worker such as a broken leg, back injury, or appendicitis. This approach meant that the worker was expected back as soon as she was physically able. More importantly, it did not consider the psychological needs of the new mother and child.

A twist to the pregnancy as "disability" definition was the recent attempt to stretch the disability designation to nursing mothers. The American College of Obstetrics and Gynecology (ACOG) was asked by some supporters of nursing mothers to include lactation as a disability. The request was made to enable mothers to nurse their babies longer before having to return to work. Currently, disability benefits for pregnancy per se normally are used up in four to six weeks. ACOG felt that this was stretching the disability concept too far. They believed the health value of nursing should be emphasized instead.

Probably the major difficulty with the disability standard is that a great number of companies do not provide any disability insurance. This is particularly true in non-union or part-time jobs where the pregnant worker has no benefits or protection at all. According to the Congressional Research Service, unions have not always been strong regarding maternity benefits and pregnant workers' rights. Maternity leave is provided for in only about 36% of collective bargaining agreements in the United States. Furthermore, provisions between union and non-union maternity leave programs do not differ markedly. In the past, unions too frequently used health and safety issues as trade-off items and reserved what power they had for negotiating better wage settlements. But what good was a better wage if you paid with your health and the health of your unborn children?

GIVING WITH ONE HAND AND TAKING AWAY WITH THE OTHER: FETAL PROTECTION POLICIES (FPPs)

The 1978 Pregnancy Disability Amendment, though flawed, genuinely attempted to protect women. Around the same time, also ostensibly designed to protect women, corporations formulated Fetal Protection Policies. What are Fetal Protection Policies? Why do they sound so good and are so bad? Fetal Protection Policies (FPPs) remove pregnant women, or sometimes all women in their childbearing years, from jobs that involve substances known or suspected of being harmful to the fetus. The catch is that this is frequently done when the substances are known or suspected to be harmful to other workers as well. Many corporations prefer the supposedly benevolent option of instituting discriminatory fetal protection policies to the far better but more expensive alternative of cleaning up the workplace for all. Too often actions taken at the individual level are employers' "cop outs" for not eliminating the *sources* of the hazards.

Fetal Protection Policies were designed to protect the employer from lawsuits rather than protect the employee or unborn child from harm. Companies feared a civil rights discrimination suit less than they did a tort action (a wrongful act which is liable under the law) for workplace-induced harm to an unborn child. Successful tort actions, though relatively rare, could result in huge monetary awards. Workers can use the torts route to seek compensation for workplace-induced reproductive health damages, since state workers' compensation laws only compensate for workplace-induced injuries that affect their ability to perform their jobs.

During the 1980s, at least 15 of the Fortune 500 largest corporations as well as numerous other hospitals and businesses were reported to have had Fetal Protection Policies. These exclusionary provisions vary greatly. Some are based on rigorous research and are carefully written and documented. Others are speculative. Some have unwritten guidelines allowing more flexibility in interpretation. The usual scenario is that you are either fired or offered substitute work at *lower wages*. This is more common in well paid traditionally male occupations. Large corporations such as American Cyanamid, Olin, and General Motors have pursued such policies.

Interestingly, FPPs were still being strongly defended by representatives of big business a decade after the general public and rank and file workers began to value strong health and safety regulations for all employees. Both of these groups understood that cleaning up the workplace costs money and they were prepared to pay the price. This is a substantial change. Prior to 1970 wages were more important than safety and health issues. By 1977, however, a third of the production workers responding to the University of Michigan Quality of Employment Survey (QES) indicated that they were willing to forgo a 10% increase in pay for "a little more" workplace safety and health. Over three-quarters of the workers in the same survey believed that they should have "a lot of" or "complete" control over safety and health.

Furthermore, in 1981 a Harris public opinion poll taken at the height of President Reagan's deregulation campaign showed that two-thirds of those polled rejected sharp cutbacks in the Occupational Safety and Health Administration's (OSHA) ability to enforce employee safety regulations.

Pregnant workers have become more active and have challenged these actions in court. Two decisions defended the right of pregnant, hospital X-ray technicians to be temporarily transferred to safer positions without losing their jobs or pay. Three Federal Appeal Courts reviewed Fetal Protection Policies in cases accusing employers of using these policies as a form of sex discrimination. All three courts ruled that it is not sex discrimination to exclude pregnant or fertile women from a particular worksite or job *only* if there is scientific evidence justifying the removal and if less discriminatory alternatives do not exist. These cases are: *Hayes v. Shelby Memorial Hospital*, 726 F.2d 1543 (11th Cir. 1984); *Zuniga v. Kleberg County Hospital*, 692 F.22d 986 (5th Cir. 1982); and *Wright v. Olin*, 697 F.2d 1172 (4th Cir. 1982). The courts stated that employers must prove that they are not discriminating against women by presenting good evidence that (1) the exposure involves a significant risk of harm to the unborn children of women employees, (2) the same level of exposure does not involve a similar risk of harm to the offspring of male employees, and (3) the FPP effectively reduces the risk. For a clear and comprehensive discussion of the arguments put forth in these judicial decisions, see Chapter 8, on "Sex Discrimination Issues," in *Reproductive Health Hazards in the Workplace*, Office of Technology Assessment, U.S. Congress. Despite these precedents and despite the fact that in 1980 a U.S. Appeals Court upheld OSHA's decision to issue standards providing the same lead safeguards for men and women, this same issue had to be fought again and again.

Most recently a district judge in Milwaukee upheld a Johnson Controls FPP barring women of childbearing age (for some bizarre reason the upper age limit was placed at 70) from any job in which their lead blood levels were likely to rise above a specific level (30ug/100g). Johnson Controls is the biggest manufacturer of automobile batteries in the United States. These areas that excluded women paid well (from $15-20 per hour) and experience in them was usually required for higher level jobs. The judge reached his decision without a trial, even though sworn expert testimony demonstrated substantial disagreement about the scientific basis for this policy. The judge interpreted prior cases to hold that a company needs only some expert testimony to support its decision. He neglected to acknowledge that it is always possible to find some expert to support a given opinion even though the vast majority of reputable scientists disagree.

The United Auto Workers appealed the lower court's decision to the Court of Appeals for the 7th Circuit. In 1989 the court issued its ruling. The majority opinion upheld the Fetal Protection Policy on the grounds that it did not violate Title VII of the Civil Rights Act.

The majority argued that the policy was not intentionally discriminatory because it was intended to benefit the offspring of both men and women workers.

Finally, the case was appealed to the United States Supreme Court. The Supreme Court accepted the appeal and heard the arguments in the fall of 1990. A friend of the court (amicus curiae) brief was written by lawyers at the Women's Rights Project of the American Civil Liberties Union Foundation and endorsed by groups of health professionals. A broad coalition of labor unions and women's groups supported the challenge to the Johnson Controls Fetal Protection Policy.

In March 1991 the Supreme Court unanimously ruled that the Johnson Controls Fetal Protection Policy violated the Federal Civil Rights Act of 1964 prohibiting sex discrimination in employment. Five Justices declared that the Civil Rights Act prohibited all Fetal Protection Policies. The four other Justices did not go that far. They believed that employers' concerns about fetal health are legitimate and that employers should be allowed to try to justify exclusionary policies. The idea that women of childbearing age should be preferentially treated out of a job on the basis of a possible future pregnancy and special harm to the unborn child does not die easily.

WHAT IT'S LIKE NOW

A pregnancy-friendly workplace is a goal that has not been reached in most Western countries. But the United States is still the only industrialized nation that does not have a national maternity or parental leave policy. This leaves pregnant workers in an ambiguous position, particularly if they work in firms that do not have clear pregnancy guidelines. Some pregnant workers do not want to tell their bosses that they are pregnant for fear of being fired. This fear can cause a considerable amount of stress, especially if the woman is the main breadwinner. Somehow, if a worker does not feel well when she is pregnant, she is considered unemployable. If she is ill and not pregnant, it is viewed as a temporary phenomenon that is not serious. This fear of being fired is often justified. Take the case of Sheila, a hotel housekeeper. She complained to her boss that she was very tired all the time and didn't have much appetite. His unsympathetic answer was, "Be glad you have a job." Marcy, whose job was soldering and wiring on an assembly line, was told by her supervisor that she would not be working there if she kept taking so much time off.

Pregnancy discrimination shows in other ways too. Sometimes you are chosen to be laid off because you are pregnant. If you protest, you are told that you were picked for some other reason.

> We had a layoff here and they chose me as the person to cut back in hours, whether I wanted it or not, because I was pregnant. Otherwise they wouldn't have. I have worked longer here than anyone else, yet my hours have been cut back to two days a week. Other people who aren't trained to do my job are trying to do it when I'm not here. That's definite discrimination.
>
> —Adrienne, sales manager
> in a manufacturing company

Furthermore, some employers stick to the letter of the law making it difficult for you to continue in your job.

> The policy of the airline is that you may continue working as a flight attendant during the first 6 months of pregnancy. They prefer that you do not stay so they make it very easy for you to leave. I left early because you still work on the airplane and they do not make any exceptions if you're not feeling well and want to miss a flight. If you are not capable of being on that flight, they say you should be on leave. They do not even make any special accommodations for maternity uniforms during this six month period.
>
> —Nora, flight attendant

At times insidious pressure is exerted.

> I think my employer is very rude to pregnant women. The administrators try to cause a lot of mental harassment because I think they are trying not to encourage women to get pregnant. When I was pregnant, they decided to eliminate one part-time job on the IV team. They did all this by talking behind

> our backs and we got all the information through rumors. When I was about 9 months pregnant, they told me that I either had to go part-time or be bumped into another job with no preparation. I think they just try to pour on the stress.
>
> The hospital has the right to have your hours changed or else they can bump you to another position. I didn't want to do either, but then the hospital offered me full-time benefits if I did three days a week. The director and assistant director of nurses lie. They say anything to your face but you can't believe them so I went to the union with this offer. The union said that this was fine but that I should get the hospital to offer this arrangement in writing. That was no big deal, but they refused to write it down so I refused to take the part-time job.
>
> —Tammy, member of an IV nursing team

THE SLOW PACE OF CHANGE

After being nudged, pushed, and pressured, businesses, unions, and the government are begrudgingly changing their attitudes and policies toward pregnant workers and working mothers. This shift is not due to altruism, but a belated realization that paid women's employment, like men's, is vital to the American economy. On an individual level, younger male managers coming up in the system are frequently members of a two career family. They too are personally confronting the issues of reproductive hazards in the workplace, pregnancy benefits, and child care.

> I'd like to see some improvement made in attitude and understanding when you have to call out because of a sick child. Now I have to lie and say I'm sick. If I call and say my son is sick, they tell me I'll have to find someone to take care of him while I work. It's hard.
>
> —Eilleen, cardiopulmonary specialist

At least some major corporations are beginning to adopt policies regarding pregnant workers, parental leaves, child care arrangements, and flexible work schedules to meet the economic and family needs of working women. For example, Levi Strauss, Corning

Glass, and Merck have instituted maternal benefit and leave policies, and Hoffmann-La Roche has been involved in child care services for its employees since 1977. In 1979 it hired its own full-time Child Care Director and opened the first corporate-sponsored child care center in New Jersey. But according to the Child Care Action Campaign, the 3500 companies having such policies make up less than a quarter of one percent of the country's six million employers.

Companies with family-oriented work policies also have a competitive advantage in recruiting. Given that the number of workers in the 20- to 24-year-old category will decline substantially over the next 10 years, many businesses will have to institute such policies if they want the best and most talented workers.

> The hospital grants a 6 week maternity leave. If you've worked there for a year, or longer, you can apply for a 4 month leave after the 6 week maternity leave. There was child care run by the hospital at another facility close by. You had to pay for it but it was very reasonable. It was just for the day, from 7-5. I think a lot of mothers that worked from 3-11 would have their kids there and their husbands would pick them up on their way home.
>
> —Clarissa, operating room nurse

Even though the record for unions nationwide does not rate an "A" for effort on women's issues, several unions have aggressively fought FPPs and supported progressive family and medical leave policies for years. Among these are the Communication Workers of America, the Service Employees International Union, 9 to 5, the United Auto Workers, the Oil, Chemical, and Atomic Workers Union, and the United Steel Workers. Even the United Mine Workers of America (UMW), which represents over 100,000 miners, mainly male and rural, unanimously adopted a proposal to make a six month unpaid parental leave part of their 1984 contract demands.

Studies and surveys in the 1980s found that the types of available maternity-related benefits in different female-dominated industries varied tremendously. Some occupations such as teaching have good maternity benefits—usually one or more years of unpaid leave written into their contracts—while women working part-time rarely have any maternity coverage at all. Small businesses, on the whole, were more likely to allow women to return to their jobs on a reduced work schedule following the birth of their child than were large companies, but they were less generous with numbers of weeks of medical leave, employer paid health insurance, or paid benefits.

In 1990 President Bush vetoed the Federal Family Leave Act, which was finally passed by the 101st Congress. The House and Senate bills contained the provision that the federal, state, and local governments and any company employing 15 or more persons must grant female and male employees up to 12 weeks of unpaid leave over a 24-month period upon the birth, adoption, or serious health condition of a child. During this period the employee would continue to contribute toward and receive health benefits and upon return to work would resume the same job or an equivalent one.

According to some family and work experts, federal and state laws providing unpaid benefits and limiting coverage to firms employing 15 or more workers would be little more than a symbolic victory for the working women most needing the benefits. These women are disproportionately employed in the small businesses exempt from coverage. Furthermore, a high percentage of women work in low-paying jobs and probably could not afford to take off 12 weeks without pay.

President Bush, however, was fearful of the financial and disruptive impact on business, despite some evidence that businesses might be financially better off giving unpaid parental leave rather than hiring and training permanent replacements. Recent surveys indicate that most companies do not seem to be hampered in their functioning by these temporary leaves, particularly when they can telephone these employees and ask questions when necessary. In 1992, Congress has again passed the Federal Family Leave Act and at this time it is not clear whether President Bush will veto it once more. Protecting Arab oil is apparently a higher priority than protecting America's families.

State legislators also have been active in this area. Twenty-eight states have passed or have in the legislative process some type of family and medical leave bill. New Jersey has one of the most comprehensive. Most of these laws apply only to large companies and many parents and potential parents are not covered. By and large, parental leave policies still have been left to private employers and their fringe benefit programs. A survey conducted by the Bureau of Labor Statistics showed that just 37% of employees in the United States are covered by a written maternity policy, and few companies have policies for other home emergencies such as caring for a sick child or parent.

It was hoped that the U.S. Supreme Court's January 1987 decision in *California Federal Savings and Loan v. Guerra*, which upheld a California statute mandating employers to provide their employees up to four months of pregnancy leave, would put pressure on other states to enact similar legislation.

The social need for quality day care is great, but in general, governmental priority for subsidizing child care programs is low. In Massachusetts, the Industrial Finance Agency, a state body that lends money to help failing industries, set aside a loan fund for companies to start child care centers at their workplaces. It was one of several agencies involved in the state-wide campaign to improve nursery care for workers' children—part of Governor Dukakis's Day Care Partnership Initiative designed to create a model day care system in cooperation with schools, businesses, and local government. The recession and the huge state budget deficit in Massachusetts squashed almost all the state's innovative efforts in this field. In New York, contracts with state employees have included child care provisions and twenty-five state workplaces offered child care services.

Former President Reagan and most other conservative Republicans have opposed federal intervention in the child care field on the grounds that it would encourage women to leave the home for work and would destroy traditional family life. During the 1980-85 period of the Reagan presidency, federal assistance to day care was cut 25%. Most of the existing assistance comes in the form of Dependent Care Tax Credits. These were formerly available for children up to 15 years of age, but the age limit has been reduced to 13 years.

Many moderate Republicans, however, believe that in the present economy support of high-quality child care facilities fosters family life rather than destroys it. For example, the price of housing has skyrocketed over the past ten years. Even the two-parent family who wants the American dream of owning a home can only afford to do so with two incomes. The altered economic conditions, combined with the rising divorce rate, have led many legislators who had formerly been opposed to day care legislation to change their minds. They no longer can pretend the need is not there. In 1970, 32% of women with children under six were in the paid work force; in 1985 the number increased to 52%. Moreover, Congressmen and Senators have begun to realize that day care is an issue that appeals to young voters who have not yet developed a strong allegiance to one political party.

In 1988 the Child Care Action Campaign, a national coalition of corporate, union, community and religious leaders, and child care experts initiated a campaign to improve child care benefits and facilities in the United States. Among the suggestions were the formation of a national child care office, the establishment of Federal regulations on minimum standards for child care, the elevation of the professional status and pay of providers of child care, and more money from the Federal Government for child care.

The 101st Congress finally enacted some child care legislation that sets up two grant programs to subsidize state child care programs. The Child Care and Development Block Grant allows the states discretion to subsidize child care for working parents from all income levels but urges that the money be directed toward low-income families. The states are required to establish a minimum quality standard. Church-sponsored day care centers will be indirectly eligible for the Federal subsidies. Parents can use vouchers if they choose day care facilities run by a religious organization. Many child care advocates are disappointed in the progress made in implementing the law. They fear that a narrow interpretation by President Bush will thwart the intent of Congress. Much to their dismay, the Bush Administration has designated an agency that handles welfare programs to administer this program rather than the agency that handles child development programs.

A much smaller program, Entitlement Funding for Child Care Services, makes it possible for states to give child care payments to families on welfare or those who would have to go on welfare because they cannot afford child care and cannot work without it. In addition, the Earned Income Tax Credit for poor working families with children was expanded.

While glacial progress is being made, the recession and our large national budget deficit make it likely that we will continue to fight an uphill battle for needed change in the immediate future. In the present fiscal climate, even very worthy new social programs are likely to be frowned upon. Furthermore, none of the initiatives taken by legislators even comes close to proposing benefits working women have won in other industrial countries. In comparison, we are offered crumbs.

In Europe, the average maternity leave is five months at full pay and was originally enacted as a form of health benefit. This was the focus in the 1970s. Since then it has been broadened to encompass support for parent-child relationships, child care, and child development, often providing paternal leave, subsidized child care, and the right to nurse the infant in the workplace.

In France, about 90% of three-year-olds spend their days in government sponsored pre-schools. France has a coordinated, comprehensive system that links child care, health care, and early education. This system is available to virtually all children and supported by virtually all political parties. Employed women in Italy receive two years' credit toward seniority when they give birth to a child. In Sweden, both parents can choose a six-hour workday until a child is eight years old. If the mother and father work different shifts, this reduces the need for child care substantially. Child care facilities, a municipal responsibility, are financed through local taxes, parents' fees, and employer payroll taxes. Home care for children who have brief illnesses is provided.

In comparison to these benefits provided by other countries, pregnant women in the United States are often fired under some pretense or rehired as new workers without seniority. Without being able to find affordable child care, many new mothers have to settle for low-paying jobs without fringe benefits because they only can work limited hours at a worksite close to home.

THE NEXT STEPS

We have come a long way since the "factory girls" labored under unsafe and unsanitary conditions for four cents an hour. But subtle types of harassment and discrimination have accompanied these improvements. These hidden agendas are more difficult to prove and in some ways more difficult to combat as the practices are defended on other grounds—efficiency, protection, equity—all of which sound laudable. Pregnancy or just the biological potential for pregnancy has been used in the past and continues to be used by employers as the unstated excuse for policies that impede equality, job security, and career mobility for women.

Compared to European policy toward work and the family, our solutions are too little and too late for the current generation of pregnant workers and working mothers with young children. While the need for maternity benefits and child care facilities remains acute, the task of achieving a workplace free of reproductive hazards remains the most difficult. Pregnant employees continue to face scientific uncertainty about reproductive risks, myths about their capabilities and place in the work force, and strong economic pressures against cleanup costs.

Much more research needs to be conducted to trace the paths of reproductive harm in both men and women and the agents that cause them. The question is what do we do in the meantime, before we have any firm answers. "Nothing" is not an acceptable answer. Knowledge is power—in the long run.

> I'd like to see personnel give any pregnant woman a list of substances that are hazardous. When I was working with about eight patients with pneumonia I was constantly bombarded with their coughing. I was sick for six weeks with a bad cough. I've been sick from chemicals they use to strip the wax off the floors and from paint fumes from the area being renovated. I'd also like them to explain all of our medical benefits to us.
>
> —Eilleen, cardiopulmonary specialist

In the short run, however, when no ready answers are available, incomplete knowledge can bring anxiety and frustration. Sometimes women prefer not to know about potential hazards they face because of the fear it brings. They prefer to pretend nothing is wrong. This is not to suggest that women panic and see reproductive hazards all around them. It does mean, though, that women have to learn to make informed individual judgments under conditions of uncertainty. They need to be able to weigh alternatives and risks in relation to their lifestyles and goals.

The following list provides ideas about how to use available information in the face of uncertainty:

- In learning to ask the right questions
- In learning to evaluate information
- In learning what we can do to protect our reproductive health both on and off the job
- In learning to put the potential workplace reproductive risk into perspective with other risks we face in our lives, e.g., driving in an automobile, smoking, living in a house with a high level of radon
- In learning not to panic about highly exaggerated stories in the media
- In choosing a healthier place to work if options are open
- In seeking improved workplace conditions
- In using occupational reproductive hazards as a focus for organizing women workers
- In obtaining information and support from unions, citizens groups, professionals, and the government
- In securing transfer of position with job security and equal pay during pregnancy

The ensuing chapters will help you evaluate information about reproductive hazards in your workplace. They clarify scientific terms and methodology, discuss legislative protection and judicial decisions, and present educational, health, community and governmental resources, as well as strategies to reduce stress related to work and family life.

SUGGESTED READINGS

Blum, M., 1983, *The Day Care Dilemma*, D.C. Heath, Lexington, MA.

Bureau of National Affairs, 1987, *Pregnancy and Employment: The Complete Handbook on Discrimination, Maternity Leave, and Health and Safety*, Rockville, MD.

Davies, M., 1978, *Maternity: Letters from Working Women*, collected by the Women's Cooperative Guild, W.W. Norton, New York.

Kamerman, S., Kahn, A. and Kingston, P. 1983, *Maternity Policies and Working Women*, Columbia University Press, New York.

Kessler-Harris, A., 1982, *Out to Work: A History of Wage-Earning Women in the United States*, Oxford University Press, New York.

Stellman, J., and Henifin, M. S., 1982, "No Fertile Women Need Apply: Employment Discrimination and Reproductive Hazards in the Workplace," in *Biological Woman – The Convenient Myth*, Hubbard, R., Henifin, M. S. and Fried, B. eds., Schenkman, Cambridge, MA.

Photo © Earl Dotter

Chapter 2

What Can Go Wrong: Reproductive Hazards for Man, Woman, and Child

LEARNING ABOUT OUR REPRODUCTIVE SELVES

Before we can improve our workplaces, we need to know a little about the biological basis of reproduction and the points of vulnerability from occupational hazards. Reproduction is a complicated business and the course of procreation is not smooth. Many fascinating pieces of the puzzle have been discovered, but still more remain to be unearthed. We particularly want to know what goes wrong and why.

About 10 to 20% of couples in their childbearing years are infertile. In about half the cases, the male is totally or partially responsible for the difficulties in conceiving. About two-thirds of infertile men suffer from some type of poor sperm function, and doctors do not know the cause. Even when conception does occur, only about one-quarter to one-third of fertilized human eggs are likely to result in the birth of a baby. Many pregnancies end in miscarriages. About three-quarters of early miscarriages show chromosomal or other abnormalities. Furthermore, about 3% of babies born in the United States have a congenital malformation (one that they are born with) and by the end of the first year another 3% are diagnosed as having some developmental defect, with additional physical and mental problems appearing later in childhood or adolescence.

Environmental influence is seen to be of major importance, and society is now putting pressure on pregnant women to adhere to certain life styles that were never questioned at earlier times. Women are told not to drink alcoholic beverages, not to smoke, to limit coffee, not to take medication unless absolutely necessary, to avoid the recreational use of drugs, and they are sometimes involuntarily removed from workplaces that employers feel may pose a threat to reproductive health.

Yet scientists still know woefully little about the reproductive effects of environmental and occupational hazards. This is particularly true concerning the vulnerability of the male partner. Until recently, researchers barely looked for damage to the male. As a result of this bias, women were, and still are, assigned the major responsibility for infertility, and any harm to their pregnancies or unborn child. This is a heavy and unwarranted burden to carry. Men, meanwhile, carry the burden of being harmed while remaining blissfully unaware of that fact.

Women, too, are relatively ignorant about male reproductive functions. We are taught from childhood that reproduction is primarily a woman's concern and accept this interpretation as a scientific fact. Even when evidence is to the contrary, women shoulder the reproductive blame at some emotional level. Interviews with infertile couples have shown that the wife frequently feels like a failure even when the biological problem lies with her husband.

Biological interpretations, however, have differed during various historical periods. These biological versions were not only based on advances made in science but on the views society held about the roles of men and women at that time. For example, in the 1600s, the male was thought to play the primary role in reproduction. As strange as it may seem today, one popular hypothesis was that each male sperm contained a complete human being. Textbook drawings showed a little person standing inside the sperm cell. The woman was merely the incubator and was seen as playing only a minor part in determining the characteristics of her child. Now, almost 400 years later the tables are turned.

WHAT CAN GO WRONG

Once we learn how complex the process of human reproduction really is, beginning with egg and sperm cell formation and culminating in the birth of a child, we may be amazed that everything turns out right the majority of the time rather than wondering what goes wrong. Known and suspected causes of infertility are shown in Figure 2.1.

FIGURE 2.1. Some Known and Suspected Causes of Infertility in Men and Women

Environmental pollutants
Workplace toxics
Infectious diseases
Genetic defects
Side-effects of medication
Inadequate nutrition
Increased age
Drugs
Stress

For instance, if the hormone balance is upset in women, ovulation may not take place. Disruption of the hormonal system in men may interfere with sperm production. Direct damage to the testes and sperm producing cells can result in insufficient or abnormal sperm. Because male and female reproductive organs develop from the same embryonic tissue and by similar developmental pathways, there may be little difference in their sensitivity to mutagens. Exposure to toxic material can reduce sexual desire and damage reproductive functions in both males and females. Such exposure can change genetic material in either the egg or sperm, resulting in a miscarriage or a live born child with birth defects. Changes in genetic material (called mutations) are passed on through generations. See Figure 2.2 for a comparison of the characteristics of egg and sperm cells.

Another area needing further research is how the body repairs genetically damaged cells from which eggs and sperm mature (precursor cells). Sometimes the body cannot repair these cells correctly. This error in the repair mechanism can cause mutations. Little is known about this process, but so far there are insufficient data to assume that male sperm precursor cells are more capable of repairing genetic damage than female egg precursor cells.

Differences Between Egg and Sperm Production

What are some of the essential differences in the way eggs and sperm are produced? Each month an ovary of a healthy woman normally releases one egg. A female baby is born with undeveloped potential eggs already in her ovaries. These are called primordial germ cells and remain dormant in what is called the oocyte stage

FIGURE 2.2. A Comparison of Female Egg Cells and Male Sperm Cells

Males	Females
1. New sperm cells are generated continuously from the age of puberty.	1. Eggs are produced from precursor cells that are are formed in the embryo.
2. Cells in testes from which sperm cells mature are present from birth and can be cumulatively exposed to mutagens.	2. Eggs mature cyclically and are released by the ovaries approximately every 28 days.
3. Sperm mature cyclically. They start developing at close intervals and run concurrently, taking 70-80 days to develop. Rapidly increasing cells are generally more sensitive to genetic damage than cells that grow slower.	3. Egg precursor cells are embedded in layers of connective tissue in the ovaries within the body.

until puberty. Sometimes it is only then that toxic harm from the prenatal stage of development becomes known. Women never generate any more oocytes and therefore a substance toxic to oocytes damages a finite supply.

While the female has a finite amount of potential eggs, they are better protected than the sperm from toxic harm because they are embedded within the ovaries. The male continues to regenerate spermatogonia (the male counterpart of the oocyte) after puberty. Sperm cells take 70 to 80 days to mature and are continuously produced from these spermatogonia. As this process takes place near the surface of the body, the developing sperm are more exposed to toxic injury than the unreleased eggs. Unless an exposure is continuous or there is chromosomal or hormonal damage affecting the growth of normal sperm, a toxic dose over a brief period of time will only hurt about a 3-month supply of sperm. Then completely fresh cells that were not exposed to the toxic substance are generated. Possibly, this is nature's way of balancing things out. The woman cannot replace her eggs once they have been damaged, but they are fairly well protected. The man can replenish his sperm supply, but it is more vulnerable to short-term toxic exposure.

The Long Road to Conception and Birth

Even if the egg and sperm remain healthy, there is still a long road ahead to conception, let alone the birth of a child. Only a few sperm ever reach the egg. For example, assume that the number of sperm reaching the ovum (egg) is probably less than 1 per 100,000 ejaculated by the male. This means that somewhere between 10 to 100 sperm actually come close to the egg. We do not know what, if anything, is special about these particular sperm. Are they just a random sample of all the sperm ejaculated at one time or are they a special group of sperm cells selected by the female's body on the basis of yet unknown criteria?

The effect of exposure to toxic substances on the fertilized egg's ability to implant itself in the womb is also unclear. Immediately after conception, any interference with the cell division of the fertilized egg or its implantation in the uterine wall causes the embryo (what the unborn child is called during the first few weeks after conception) to die. The ensuing miscarriage is so early that women only realize that their menstrual period may be a few days late or heavier than usual. If the embryo survives the early days it has passed its first hurdle, but there are more to come.

The embryo, and later the fetus, is usually more sensitive to toxic substances than the mother. This can lead to problems since most substances present in the mother's bloodstream, whether they entered through her digestive system, were inhaled, or absorbed through the skin, pass through the placenta to her unborn child. The first three months of pregnancy are an extremely sensitive period because during this time the organs are formed and developed, and major deformities in the unborn baby's heart, brain, limbs, or other organs can occur. Substances causing these types of defects are called teratogens. They can cause devastating fetal damage during a specific pregnancy but are neither transmitted to future generations nor likely to occur in another pregnancy unless the mother is again exposed to the hazardous substance.

During the second and third trimesters the fetus is developing and growing. Childhood developmental problems can originate during this period. Anything that causes the baby to be born prematurely or have a low birth weight, e.g., stress, toxic exposures, or poor nutrition, puts it at more risk for illness and death. Table 2.1 lists ways of checking female and male reproductive health.

TABLE 2.1
Checking Reproductive Health

Female Health

1. Personal, medical, and family history
2. Physical examination
3. Functioning of the ovaries
4. Secondary sex characteristics
5. Receptivity to sperm of cervical mucous
6. Monthly thickening of the uterine wall (endometrial cells)
7. Condition of fallopian tubes and uterus

Male Health

1. Personal, medical, and family history
2. Physical examination
3. Semen quality
   a. Appearance of semen
   b. Ph (degree of acidity) of semen
   c. Amount of semen ejaculated
   d. Number of sperm per milliliter of semen
   e. Sperm movement patterns
   f. Ration of live/dead sperm in semen
   g. Sperm shape and size
   h. Ability to penetrate cervical mucous
   i. Sperm-egg interaction

Adapted from material in U.S. Congress, Office of Technology Assessment, *Reproductive Health Hazards in the Workplace*, U.S. Government Printing Office, Washington, December 1985, Chapter 5.

As we have seen, reproductive problems occur at every step of the process, and as a result, toxic substances can cause reproductive harm at all stages—prior to conception, during pregnancy, and (as will be discussed at the end of this chapter) after birth. Toxics can affect menstruation, ovulation, sperm production, and sperm quality, as well as cause miscarriages, birth defects and cancer.

BRINGING THE MALE BACK IN

Prospective parents can find many books and articles about women and what can go wrong with pregnancy, but few have been written about men and what can go wrong with pregnancy. Not only has the "little person standing inside the sperm cell" disappeared from these books, but so, too often, has the big person who is generating the sperm. Because the emphasis in reproductive research has focused on the woman's role, the potential for birth defects arising from the result of the father's exposure to chemical substances is still largely unknown. New tools in molecular biology may change this. These tools identify mechanisms that could link toxic exposure to sperm damage which in turn could cause birth defects. Scientists can now examine hundreds of newly discovered proteins in sperm, place special markers on them, and watch how chemicals interact with these proteins and DNA.

On the basis of current understanding of human reproduction, birth defects are thought to occur in three ways: (1) by gene mutation, (2) by chromosomal mutation, or (3) by acting as a teratogen (interfering with normal embryonic or fetal development) carried in the semen during intercourse.

Some scientists think that toxic substances absorbed by the exposed male may contaminate the seminal fluid and cross the placental barrier through intercourse. They then can be absorbed by the fertilized egg causing miscarriages or birth defects. A few studies and case reports suggest that this may indeed happen. One report describes a woman who bore three children with severe malformations over a period of years when her husband was suffering from lead poisoning. Subsequently all three children died. Pregnancies before her husband had developed lead poisoning and after he had recovered led to the birth of normal children.

Toxics and the Male

As far back as 1934, Dr. Alice Hamilton believed that lead could affect sperm cells. She based this opinion on a Japanese study reporting that twice as many couples where the husband had been exposed to lead were childless than would be expected. A 1985 review of scientific papers analyzing the effects of lead on human health also revealed an unusually high number of abnormal pregnancies among wives of lead workers. Although there is increased evidence that these abnormalities are likely due to the male's exposure to lead, so far there have been no *conclusive* studies proving or disproving this.

Male employees exposed to anesthetic gases have also been studied and the findings have differed. Some studies have found higher rates of miscarriages among partners of exposed male operating room personnel and dentists while others did not. Other studies discerned a small increase in some types of birth defects in the offspring of these exposed men. One of the problems posed by investigations based on workers' memories of past events (retrospective studies) is that there may be biased recall of both the extent of exposure and exact reproductive injury.

The association between reproductive harm and male toxic exposure also shows up in animal studies. These studies have identified more than 100 chemicals that produced spontaneous abortion or birth defects in offspring of exposed males, and the list of agents suspected of causing reproductive damage through human male exposure keeps getting longer. Men who work in factories that produce plastic products, and fire fighters who come into contact with burning plastic material often inhale vinyl chloride (VC). Vinyl chloride can affect genetic material in sperm. Studies in four countries have found that workers exposed to polyvinyl chloride (PVC), one of the most commonly used plastics, have an excess of chromosomal changes. Wives of workers exposed to vinyl chloride have been found to have higher than expected miscarriage and stillbirth rates.

Still More Toxic Villains

Ionizing radiation, carbon disulfide, estrogen, DBCP, and extreme heat are other villains. Ionizing radiation can kill sperm-producing cells of the male testes. A study of Japanese radiological technicians showed that both males and females had a higher sterility rate than the general population. Research on male workers exposed to carbon disulfide for three years (used as an insecticide, solvent, and in the manufacture of viscose rayon) found that the exposed men had decreased sex drive and trouble in having an erection. They also had five times greater rates of sperm abnormalities than the control group not exposed to the chemical.

Several pesticides and herbicides have been linked to male reproductive organ dysfunction and to birth defects in the offspring of exposed male workers. In one report four out of five members of a farm working crew exposed periodically to a wide variety of pesticides complained of impotence. Accumulating data revealing that Dinoseb, which had been sprayed on crops in the United States for more than 40 years, could cause birth defects, and sterility, as well as cancer, cataracts, and immune system damage, finally became persuasive enough for the EPA to use its emergency powers to order an immediate ban in October 1986. The only two other times that the EPA used its emergency powers was to ban the agricultural chemicals 2,4,5 -t, an ingredient of Agent Orange, and ethylene dibromide (EDB), a strong human carcinogen that was suspected of damaging both the quality and quantity of exposed men's sperm.

Paternal exposure to electromagnetic energy is also suspected as being a contributing factor to their offspring being born with either a clubfoot or Down's syndrome. In addition, mutagenic and carcinogenic effects have been reported. This is quite worrisome, as advances in magnetic and superconductor technology mean that even more workers will be exposed to weak electromagnetic fields. Researchers in Europe and the United States are currently investigating the possible links between electromagnetic fields and harm to human reproduction.

To further add to the complexity, some substances can have several effects. For example, the pesticide DBCP is a mutagen that can cause direct damage to the chromosomes. It can also destroy the cells in the testes that are responsible for producing sperm (the male germinal epithelium) and thus cause male sterility. In 1977 semen samples collected from 308 workers exposed to DBCP found that 50% of these men either had no sperm or a very low sperm count in their semen, nearly twice the rate expected for the general population. Impotence, as well as sensitive and enlarged breasts, are symptoms of men who produce estrogens, estrogen-based drugs, DES pellets and pastes, as well as men who administer DES to animals.

Figure 2.3 shows work exposures associated with abnormal human male functioning and Table 2.2 lists substances harming males or their offspring.

FIGURE 2.3. Work Exposures Associated with Abnormal Human Male Function

Confirmed Effect	Inconclusive Effects	No Observed Effects
carbon disulfide	anesthetics	epichloro-hydrin
DBCP	arsenic	gylcerine
lead	benzene	P-TBBA
oral contraceptives	boron	PBB
	cadmium	PCP
	carbaryl chlor-decone (kepone)	
	DNT and TDA	
	ethylene dibromide	
	manganese	
	mercury	
	pesticides	
	PCP	
	radiation, ionizing	
	radiation, non-ionizing	
	solvents	
	TCDD (dioxin)	
	vinyl chloride	

Adapted from Schrag, Susan D., and Robert Dixon in WOHRC News, vol. 8, no. 2.1, Women's Occupational Health Resource Center, School of Public Health, Columbia University, New York, 1986.

TABLE 2.2. Some Substances Known or Suspected of Harming Male Reproductive Health or the Health of Their Offspring

1. *Lead used in making storage batteries and paints:* Fewer sperm, sperm that moved more slowly than normal (decreased sperm motility), and more abnormal-shaped sperm (increased malformation of sperm).

2. *DBCP (dibromochloropropane) a soil fumigant now banned:* Mutagen, lowered sperm count, testicular dysfunction.

3. *Ionizing and non-ionizing radiation, the former found in nuclear plants and medical facilities, the latter in high voltage switchyards and in communications facilities:* Possible damage to sperm cells and lowered fertility.

4. *Anesthetic gases:* Unexposed female partners are thought to have higher than normal number of miscarriages.

5. *Vinyl chloride used in plastic manufacturing:* Unexposed female partners are thought to have more miscarriages and stillbirths.
6. *Kepone used as a pesticide:* Possible loss of sex drive, lowered sperm count and slower movement of sperm.
7. *Heat stress occurring in foundries, smelters, bakeries and farm work:* Lower sperm counts and sterility.
8. *Carbon disulfide used in the manufacture of viscose rayon and as a fumigant:* Possible loss of sex drive, impotence, and abnormal sperm.
9. *Estrogen used in manufacturing of oral contraceptives:* Possible loss of sex drive and enlarged and sore breasts.
10. *Methylene chloride used as a solvent in paint strippers:* Possible very low sperm counts and shrunken testicles.
11. *EDB (ethylene dibromide) used as an ingredient in leaded gasoline and as a fumigant on tropical fruit for export:* Possible lower sperm count and decreased fertility in wives of workers.

In the Harm of Heat's Way

Because so little emphasis has been placed on male reproductive hazards for so long, both the development of research studies in this area and the distribution of new research findings seem ever so slow. Even when an agent, or production process such as one involving extreme heat is suspected of being a reproductive hazard, this information is not always forwarded to the employer, and even when the employer knows, he may deliberately neglect to pass this information on to his employees on the grounds that it has not yet been proved.

Take the case of Ken who was a forker in a heating and plumbing manufacturing firm. It was his job to fire pieces in the furnace to prepare them for enamel finishing. His employer never informed the men working in the furnace that the intense heat could damage their sperm count. It was only when it was too late that Ken learned the truth.

> I didn't realize there was a problem until the doctor told me my sperm count was too low to impregnate my wife. He told me that he was 90% sure my sperm count was irreversibly low because of the intense heat at my job. There is no medical

treatment available to correct my problem. I was angry enough to take the company to court, but I didn't because my doctor told me it would be too hard to prove my low sperm count was caused by the heat work. He said they could say that taking hot showers also caused low sperm counts.

—Ken, furnace forker in a heating and plumbing manufacturing firm

Two other forkers that Ken knew about also had fertility problems. One was permanently sterile and the other, who had only been on the job for a year, eventually had children after transferring to another department. Despite these known cases, the company has never changed working conditions to reduce the potential risk of infertility nor informed workers of this possibility.

In Ken's case, the probable cause of his infertility was easy to determine. But long time lags between exposure and reproductive damage sometimes make it difficult to establish cause and effect.

MEN AND THEIR PERCEIVED REPRODUCTIVE VULNERABILITY

While men are very aware of the risk of developing cancer from occupational toxics, they focus less on reproductive harm despite evidence linking their fertility and their partners' miscarriages to their own workplace exposure. Part of the denial of a male role is due to society's narrow interpretation of reproductive harm—one limited to the effects on the pregnant woman and unborn child. The macho image also operates. Generally, men do not like to talk about any difficulties remotely concerned with sexual adequacy and prowess. When asked about whether he ever lost his sex drive, became impotent, or had a low sperm count, one of the men interviewed responded:

Now look, between my wife's headaches and backaches and having to get up early and things like that you can't afford to lose no sex drive. You better get it while it's hot.

—Robert, plastics fabricator at an aircraft plant

Even when men are aware of the hazards, they do not seem to vocalize their concerns as much as women do. A few interviews showed that men, more than women, are likely to accept the reassurances of their companies and unions. Take the case of Alan, a firefighter. Firefighting is one of the most hazardous jobs in terms of exposure to all sorts of highly dangerous toxics. Despite the progress firefighters' unions have made in providing information, firefighters still do not know what hazardous substances they are facing in most fires. Yet Alan, outwardly at least, remains reassured:

> In any fire there will be gases from the smoke. When different plastics burn they give off poisonous smoke. It all depends on who has the fire. The closest I came in contact with poisonous gases was in one house fire where there was a refrigerator smoldering. I guess it was insulation on the inside so we pushed that outside the door without our masks on and all of us were sick for a week.
>
> As long as we're aware of what is on the premises, we can do everything possible to prevent any harm to our offspring from our exposure. But as long as we're not sure what different chemicals are in the building, there is no way for us to be totally protected. The fireman's national organization is always doing something. There is always someone in the national, state, or county federations who is working on the problem and trying to get everything that we need to make our job much safer. Unlike a lot of jobs, information doesn't filter down as slowly with us because we're a national organization and the information is usually distributed rather quickly. How fast somebody might act on it from city to city may vary and getting the right equipment takes more time and money.
>
> I am under a lot of stress in my job and at times my sex drive is increased. I think it is primarily to relieve tension, because afterward you feel rather relaxed. You know how it is, don't you?
>
> —Alan, firefighter

Male workers operating in other types of toxic environments also voice concern that is combined with either a general satisfaction with safety precautions or a refusal to think about the implications of the hazards.

> We come in contact with a lot chemicals such as ink solvents and resins. Resins are a powdery substance that would probably have a danger of going into your lungs. We also use some acids and a couple of dangerous gases – hydrogen, nitrogen, and things like that. But we use a lot of precautions.
>
> I had an opportunity to work with machinery that has radiation, but I was not exposed to it. There was a chance of inhaling the gases. I never inhaled it. We work for a union company and everyone is concerned with strict safety. We work with a mask, proper equipment, goggles, and gloves. You work with these substances but don't come in contact with them. They have safety inspectors come in periodically to see if everybody is working safely. It is very dangerous, a lot of bad things could happen.
>
> I can see something happening to myself or my children if sometimes I got careless or I got into a situation where I was exposed to the chemicals. I am very conscious of that so I take precautions. I don't see it really happening.
>
> – Larry, lab technician in a linoleum factory

> We smooth out plastic parts to be put in the inside of airplanes. Sometimes we paint them. I come in contact with different paints and there is dust all over from sanding the plastic. Sometimes I'm pulled out of where I'm working and have to go downstairs where they are doing all the spray painting. Even to go to the bathroom, you have to go through the painting section. Downstairs the air was better because they had the doors open. Upstairs there were only a few windows. Over the last year or so I've become more conscious of the dust in the air. I figure it's uncomfortable thinking about it so I turn off about it. But the safety factor keeps sticking in my head. It's hard to say in a factory like this what the other guys and their families think. Guys take off sick and they may go fishing or something like that. We're a close group of guys, but we don't really know anything about each other. My company is a good

company. I'm not trying to bad-mouth the company. But I figure the dust is just something that nobody paid too much attention to.

—Robert, plastics fabricator
in an aircraft plant

The male workers expressing their concern try to be extra careful and tend to feel successful in their efforts, whereas the women remain more uncertain.

A LINK BETWEEN CANCER AND REPRODUCTIVE HEALTH?

In addition to specific male and female reproductive hazards, scientists worry about workplace substances that might cause cancer of the reproductive organs or cancer in the offspring. Will reducing workplace exposure to the risk of cancer also help eliminate reproductive risks? Researchers believe that there is a large overlap in the substances that cause mutations and those that cause cancer. Most carcinogens (agents that cause cancer) are mutagens (agents that change genetic material) and some mutagens are also carcinogens. If an agent is a known mutagen or carcinogen, it also may be a reproductive hazard. Because of this overlap, some scientists feel that until improved techniques are developed for detecting mutations directly, one way to eliminate them is by having a strong policy protecting individuals against carcinogens.

Cancer experts estimate that 80 to 90% of all human cancers might be linked to exposure to food additives, drugs, radiation, industrial chemicals, and smoking. Many food additives that can cause cancer have been taken off the market and cigarettes have health warnings printed on the packages, but a large percentage of industrial chemicals have never been tested to determine if they are cancer-causing. Others, like formaldehyde, which has caused cancer in laboratory animals and has been linked to illnesses in humans, are still being used widely. Regulating agencies believe there is insufficient scientific evidence relating low exposure to disease in humans.

Cancer is known to interfere with conception itself when it affects male and female reproductive organs. Men who have had frequent skin contact with cutting and lubricating oils on their jobs have developed cancers of the scrotum, and male rubber workers seem to have higher than expected rates of cancer of the prostate. In addition, though very little research has been conducted in this area, a few studies seem to indicate a link between childhood cancer and the father's exposure to hydrocarbons and lead prior to conception.

Female cancer can involve the cervix, uterus, or ovaries. Physicians advise cancer patients of both sexes who are undergoing radiation and chemotherapy not to attempt to conceive children during this time as these treatments are extremely hazardous to egg and sperm cells and to the developing fetus. Health care workers handling chemotherapy or who work in X-ray and nuclear medicine departments should exert caution as well. Some cancer-causing agents can cross the placenta, thus causing direct reproductive damage to the fetus. While the placenta acts as a partial barrier, it by no means filters out all potential toxics. Because the lag between exposure to a carcinogen and the appearance of cancer can range from 5 to 40 years, many children who were exposed as fetuses may develop cancer many years later and they and their parents never realize the connection.

PRENATAL DIAGNOSIS AND PREGNANCY HAZARD COUNSELING

While most obstetricians and nurse-midwives are usually aware of recent developments in prenatal diagnostic testing and infertility treatments, few have much knowledge about occupational and environmental reproductive hazards. Some do not even know sources of referral. As a response to requests by pregnant women and their spouses for more information about the effects of workplace and environmental exposures, smoking, alcohol, and drugs, pregnancy hotlines have been established. The hotlines are free, but the in-person consultations are not. There are now 18 throughout the country and more are being planned. Telephone numbers and addresses are given at the back of this chapter. Many are located in genetic counseling clinics of large university medical centers.

If you were to call, you would be asked questions not only about your exposure to the suspected hazardous substance but about your family, work, and health histories and other pertinent information. Usually the pregnancy hotline can provide answers on the phone. This is particularly true if you have common questions concerning smoking, drinking, and taking drugs during pregnancy.

Your fears about the impact of occupational and environmental exposures, an area in which firm data is more limited, are more difficult to allay and may require further investigation and an in-person consultation. If there is insufficient evidence to determine reproductive harm, the telephone counselor might reassure you that so far there is no proof that the substances you are exposed to pose a substantial pregnancy risk. She would suggest that, given the lack of sufficient information, you should take precautions in order to keep high workplace exposures to a minimum. These precautions involve very simple things such as washing hands more frequently, not eating or drinking while working with a suspected chemical hazard, or shutting off the VDT when it is not in use.

If you ask about a particularly complex or rare exposure, the hotline counselor will research the problem and call you back or ask you to come in for a consultation. If you have been exposed to an agent or substance that is known or suspected to cause reproductive harm, e.g., X rays, lead, mercury, ethylene oxide, you are almost always asked to come in for a personal counseling session. The counselor will explain the extent of the risk you face, the kind of damage to your unborn child that might occur, and whether any of the prenatal diagnostic tests would be of use in determining whether the fetus was affected. Such counseling sessions and prenatal diagnosis are often covered by health insurance. The three main prenatal diagnostic techniques – ultrasound, amniocentesis, and chorionic villus sampling (CVS) – are briefly described at the end of the chapter.

Pre-Pregnancy Counseling

Rather than wait until a couple conceives, some reproductive toxicologists are now considering recommending pre-pregnancy counseling for couples, especially those who may be at particular risk because of a family history of a genetic disease or occupational or environmental exposures.

When a couple plans to have a child in the near future, they make an appointment with a pre-pregnancy counselor who explains life-style, genetic, workplace, and environmental risks. This type of counseling would be considered the first step in obtaining good pre-natal care. Armed with the information, couples could decide what life style and occupational changes to make. These decisions are particularly important for two reasons. First, the effect of harmful exposures on the potential father's sperm in the weeks and possibly months before conception can affect the health of his future child. Second, the unborn child is particularly susceptible to harm early in pregnancy. Now early prenatal care means seeing your doctor or midwife during the first month or two of pregnancy. In the future, early care might encompass pre-conception counseling as well.

While both pre-conception counseling and prenatal diagnostic techniques are helpful, they do nothing to alleviate occupational or environmental conditions causing the possible danger in the first place. Companies' educational health improvement programs for their pregnant workers can be used as a substitute for cleaning up the workplace. This is like locking the barn door after the horse has been stolen. With the current emphasis on individual responsibility for one's own health, identifying the occupational and environmental culprits takes last place. It's no mystery why this is so—the hazardous conditions are frequently expensive and difficult to correct. It is easier and cheaper to provide a pregnant woman with health information and diagnostic technology and then blame her for not using them if anything goes wrong.

Currently, prenatal diagnostic techniques cannot detect the effect of most occupational and environmental exposures. Not enough is known about what to look for. For example, most scientists believe that exposure to occupational and environmental toxic substances are more likely to cause miscarriages, or developmental problems in the child, rather than birth defects per se. So many factors can cause these occurrences that it is extremely difficult to pin the blame on any one or even combination of substances. As more data are obtained, this picture will change.

Promising research is being conducted on the harmful interaction of certain genetic characteristics and particular industrial chemicals. Carrier and prenatal diagnostic tests for "susceptible" workers are likely to follow. These breakthroughs will then likely lead to the identification and counseling of women who are genetically more susceptible to specific industrial chemicals. Atypical reactions to drugs such as barbiturates and halothane have already been found in individuals with certain genetic disorders and in people with certain normal genetic variations.

Genetic variability can offer protection as well as susceptibility to industrial chemicals. One hypothesis is that the workers' genetic makeups account for the fact that under the same working conditions some workers develop a certain disease while others do not. This kind of knowledge can lead to the prevention of reproductive damage as well, but raises serious ethical, social, and economic questions. Continuing breakthroughs in molecular biology will enable scientists to identify additional genes that may be responsible for harmful interactions between our bodies and our environment. Women need to be alert to this direction of scientific research, since they are likely to be eventual recipients of new genetic technology that will be developed. (See section on genetic testing at the end of the chapter.)

TOXIC EXPOSURE AFTER BIRTH

In addition to increased awareness about male reproductive workplace hazards, recognition of workplace hazards affecting infants and children is growing. Just when you are breathing a sigh of relief that you had a healthy pregnancy and that your baby was born without any birth defects, new worries emerge. The two most common avenues of toxic exposure to a child after it is born are from dust brought home on the parents' work clothes and through breast milk. For example, because estrogen can be absorbed through the skin or inhaled, young sons and daughters exposed to estrogen dust brought home on their mothers' or fathers' clothing develop sore and enlarged breasts which disappear when exposure to estrogen ceases.

Men working with asbestos bring home asbestos dust on their clothing, affecting their wives' and children's health. In Manville, New Jersey (home of the Johns Manville company, which claimed bankruptcy rather than be liable for millions of dollars of health damage claims against the company), research physicians are studying the possible health effects on the third generation—the grandchildren of the exposed employees.

Breast milk pollution is a newer problem. Recently, the World Health Organization presented worrisome evidence that toxics, entering mainly through the route of the food chain, are now found in women's breast milk in most areas of the world. How serious is the problem? Does the infant suffer any ill effects?

Breast Milk Pollution

A nursing mother typically produces a liter of milk per day providing her baby with nature's best nourishment. Unfortunately in today's contaminated atmosphere, breast milk can also pass on a wide variety of substances present in the mother's tissues or blood. Chemicals or drugs can bind to milk protein or to the surface of milk fat globules and are included in the breast milk along with protein, fat, carbohydrates, minerals, vitamins, hormones, and antibodies.

Currently there are no internationally accepted figures for safe levels of breast milk pollutants. Until there are, *doctors recommend that you continue breast-feeding your child* except for cases of extreme contamination. The advantages of breast milk are still seen to be overwhelming because of the protection it provides against infection and the strengthening of the immune system.

Scientists studying the breast milk path of possible health effects on the nursing infant assume the existence of a dose response. This means that a high concentration of a toxic chemical found in breast milk might harm the baby but a small amount of the same toxic chemical probably would not. They are looking into acute short-term effects, long-term chronic effects due to the build up of small amounts of toxic agents over time, and the interactive effect between chemicals. Sometimes a chemical is not harmful by itself but becomes harmful if another chemical that it interacts with is also present.

Accidental "Spills" and Breast Milk Pollution

Toxics in the workplace have not sparked concern about breast milk pollution, but several accidental chemical spills in several countries have. (Discovering the HIV virus in breast milk was a further disturbing finding.) In Japan, a faulty production process resulted in PCBs getting into rice cooking oil. Some unborn babies were exposed while in the uterus and this was further complicated by PCBs reaching the baby through their mother's milk. Similar incidents occurred involving acute methyl mercury poisoning. People ate contaminated fish in Japan and contaminated grain in Iraq. One of the consequences was that pregnant women and nursing mothers developed high mercury blood and milk concentrates, which affected fetuses and nursing infants, leaving them with damaged central nervous systems and suffering from a disease resembling cerebral palsy.

In Michigan, PCB-contaminated fish and cows fed with PBB-contaminated grain resulted in high levels of these chemicals being present in the food chain. Even though high amounts of PCBs and PBBs were found in the milk of nursing mothers, their babies did not seem to suffer any ill effects. In Hawaii, the pesticide heptachlor sprayed on pineapple tops (which had been taken off the market for all other purposes because it was considered to be too toxic to humans) also found its way into the food chain and eventually into breast milk and the nursing infants. This raised quite an uproar. Heptachlor was finally banned from use on the Hawaiian pineapple crop.

Substances to Avoid

Ideally if you plan to nurse your baby you should not be exposed to mercury, lead, solvents, many pesticides, or PCBs and PBBs in the workplace before you attempt to conceive, during pregnancy, or during lactation. You should also try to avoid exposure to anesthetic agents, cadmium, and pharmaceutical agents, particularly estrogens and chemicals involved in the production of viscose rayon and synthetic rubber. If you are self-employed, work as household help, use chemicals in hobbies such as arts and crafts, or engage in home improvements, you should also be alert to possible contaminants of your breast milk.

Contrary to what you might think, your past maternal exposure produces a body burden which you cannot readily excrete in feces or urine. Instead, chemicals retained in your tissues and bones are released when you begin to nurse. Chemicals that occur at higher concentrations in milk than in your blood plasma are the most worrisome. Even if the transfer of the chemicals from your blood to your milk is extremely low, a high concentration can accrue in the milk because the blood flow to the mammary tissue exceeds milk production 400 to 500 times.

The Lactation Process

During lactation, transfer to milk occurs readily for fat soluble substances due to the high fat content of breast milk. It is likely that fat-soluble chemicals such as DDT, PCB, and many other pesticides may be trapped entirely within the milk fat globule – not a pleasant thought. The amount of PCBs, heptachlor, and other similar toxics in the mother's body is determined by her long-term exposure. For example, the pesticide DDT has been banned by the FDA for years, yet it is still present in the food chain. There is, therefore, no dietary program that you can follow during pregnancy and lactation that will eliminate these hazards, though changes may reduce current exposure. Moreover, there are no specific tests to determine the possible effects on your nursing baby. These chemicals are difficult to metabolize or release except in the milk of nursing mothers. PCB levels, though, seem to diminish over the course of breastfeeding and number of pregnancies. So if you have many children (which is not very typical anymore) and breast-feed for a long period of time, your younger children may receive fewer PCBs in their milk.

So far there have been no documented cases of babies being harmed from being nursed by mothers occupationally exposed to chemical toxics, though blood concentrations of toxic chemicals in occupationally exposed workers typically exceed those in the normal population from 10 to 300 times. There is, however, a recorded incident involving a mother who regularly nursed her infant while visiting her husband in a dry cleaning establishment. Her baby developed jaundice, as her breast milk was found to contain perchlorethylene.

Some health experts suggest that women with pesticide or halogenated hydrocarbon solvent residues above the 90 percentile for the general population, or with mercury levels in their milk above 4 pg/liter (the measurement scientists use), or levels in their blood over 20 pg/liter should not nurse their children. After reviewing the available evidence, however, the North Carolina breast milk and formula project found that though exposure of the fetus in the womb to many of these substances is less than breast milk exposure, breast-fed infants usually thrive. Therefore, the project directors advised against *mass screening* of women's milk. For individual cases, the advice may be different. If you believe that you have been subject to excessive exposure to any of the toxic chemicals discussed, you would be prudent to have your breast milk tested.

You should be aware of the well-founded concern about increases in the total toxic load carried by women's bodies, but you should not be unduly worried about nursing unless you think you are occupationally exposed to high levels of any chemicals that have the following characteristics:

1. Not water soluble (PCBs and PBBs)
2. Not readily metabolized by your body (lead)
3. Stored in your body and slow to clear from your milk (lead, heptachlor, DDT)

If you are not sure about your workplace exposure and remain apprehensive, find out the chemical names of the substances that you are working with and tell your health care provider. Ask your doctor or nurse-midwife to check with the state environmental health department or occupational health division of a medical school for the latest research findings in this rapidly developing area of research.

WHO STANDARDS FOR CHEMICALS IN FOOD

The World Health Organization (WHO) and other agencies have published guidelines for safe chemicals in food based on studies of animal toxicity and from reported poisoning episodes in humans. These standards have been developed for adults. Infants may be more or less susceptible to specific chemicals. Nobody knows for sure.

The two standards used are ADI (allowable daily intake) and MAC (maximum allowable concentration). The ADI refers to the daily dose that over a lifetime appears to cause no appreciable risk. These dietary estimates were determined with the knowledge that such chemicals accumulate in the body, and the authors attempted to incorporate a margin of safety. So far the margin of safety of the ADIs are not known, nor can definite upper danger limits be specified. At best, you can use the ADIs as rough guides. The same advice can be given for MACs. These maximum allowable concentrations were designed to incorporate a 10-fold safety factor.

In the two major cases of organic mercury poisoning in Japan and Iraq, the MAC seemed fairly accurate. The maternal blood and milk concentrations as well as the nursing infants' blood levels of mercury all exceeded the MAC 10 to 100 times, which is above the 10-fold safety factor.

WHAT CAN YOU DO?

The following suggestions may help:

1. Avoid eating fish found in polluted waters. Check with your local and state health department or the Environmental Protection Agency (EPA) to find out which fish are considered safe to eat regularly, occasionally, or to avoid altogether.

2. Avoid quick weight loss as this will release the toxics into your bloodstream and then into your breast milk more rapidly. Do not embark on a rigorous diet until after you have stopped nursing.

3. Change your diet. Eat less meat, dairy and fish. Chlorinated hydrocarbons are more fat-soluble than water-soluble and therefore tend to remain in the food chain, especially in meat, dairy, and fish products. Cutting down on these types of foods will reduce your

current exposure but will not have an effect on the cumulative toxics already in your system.

4. As a very last resort, change jobs if you are occupationally exposed. Electrical, chemical or agricultural work, and auto industries are particularly hazardous.

5. When you are renting or buying a home, try to avoid moving into one that is located near a waste dump or incinerator.

6. Toxics in breast milk is a new issue of concern that requires a great deal more research and greater dissemination of the data already available. Organize around the breast milk contamination problem. Persuade the government to fund further research in order to establish guidelines for estimating the point, if any, where the extent of contamination of breast milk outweighs the benefit. Breast milk pollution can also be used as a rallying point for instituting public cleanup campaigns.

WHERE WE STAND NOW

Preliminary findings indicate that *parental* (both mother and father) exposure to environmental and occupational toxics is a prime suspect in causing some cases of infertility, miscarriages, reproductive abnormality of both sexes, loss of sex drive, menstrual irregularities, lack of ovulation, male impotence, birth defects, childhood and adult cancers, mental and physical developmental problems, and hereditary defects that may only appear in future generations.

However, the evidence is not in on many issues that are of concern to the pregnant worker. For example, how much of a hazard, if any, are known toxics in breast milk to breast-feeding infants? If they are, which toxics are the most harmful and at what level will harm occur? Will occupational exposure be likely to be severe enough for this dangerous level to be reached? Will general environmental exposure become so high that a small additional exposure in the workplace is enough to tip the scale? To date there has been no absolutely conclusive proof, only strong suspicions, that men exposed to occupational reproductive hazards can pass the effects of this exposure directly to their offspring. These suspicions have been bolstered, however, by evidence that the risk for fathering babies with a birth defect caused by a fresh dominant mutation (one that will show an effect even if the child inherits one normal gene) increases with advancing paternal age.

Formerly, it was thought that advancing age only affected egg cells, not sperm cells. It was assumed that if an additional or abnormal chromosome was found it came from the mother. Now studies show that between 5% and 25% of babies born with Down's syndrome receive the extra chromosome that causes this form of mental retardation from their fathers. In general, as people get older they have been exposed to more toxic substances in the workplace, at home, and in the community, and these exposures may be responsible for at least part of the association found between advanced paternal age and Down's syndrome. (Information on genetic testing and pregnancy hazard hotlines is given at the end of the chapter.)

Many women born in the baby boom of the post World War II years are already postponing having children until they are in their thirties. They cannot postpone their childbearing long enough for definitive evidence to be gathered about male reproductive hazards in the workplace or breast milk pollution. They have to protect themselves as well as they can based on the available data. Blaming the victim, however, by trying to assign responsibility to the female or male partner, is the wrong way of looking at risks to reproductive health. Expending energy at cleaning up your workplace and environment for *both sexes* is far more productive. Chapters 3, 4, and 5 will go into suspected reproductive hazards of workplaces and the home in greater detail and suggest ways of reducing them. Many women have taken the lead in this area and some of their initiatives are described in Chapter 7.

THE THREE MAIN PRENATAL DIAGNOSTIC TECHNIQUES

Ultrasound

Ultrasound analyzes embryo/fetal development. It relies on differences in acoustic densities to form images of the unborn child. During the second trimester, it can detect major fetal malformations (such as loss of limbs in the case of the thalidomide babies), multiple pregnancies, and progress of fetal growth. In later stages it can monitor fetal breathing, trunk and limb movement, and quantity of amniotic fluid.

There are three levels of ultrasound. Level 1 provides a general picture of the fetus—its overall well-being and gestational age. Level 2 is more sophisticated and looks at specific birth defects such as hydrocephaly (brain filled with fluid), microcephaly (a smaller than normal size brain), neural tube and heart defects. Though birth defects are rare, these are the more common ones and could be related to occupational and environmental exposure. Thus level 2 ultrasound can be used to diagnose such problems. Level 3 ultrasound is used for fetal surgery.

Amniocentesis

Amniocentesis consists of inserting a needle through the mother's abdomen into the amniotic cavity about the 16th week of pregnancy and extracting amniotic fluid. The amniotic sac is the fluid-filled cavity that surrounds the developing fetus. This amniotic fluid contains some live cells shed by the fetus. The fluid itself and the cells within it can be analyzed for important diagnostic information about the fetus. Until recently it was thought that the test could not be conducted earlier because there were not as yet enough fetal cells to be cultured. Now under the guidance of ultrasound, amniocentesis is being successfully carried out at 12 to 14 weeks of pregnancy at special investigative centers.

Amniocentesis is used mainly to diagnose chromosomal abnormalities like Down's syndrome and other genetic disorders. The amniotic fluid can also be tested for certain enzyme and protein abnormalities. The most common analysis is for alpha fetoprotein (AFP). Abnormally high levels of AFP are associated with neural tube defects like spina bifida (opening in the spinal column) and anencephaly (a pinhead size brain) and unusually low levels are associated with Down's syndrome.

Currently, pregnant women are routinely offered a maternal screening blood test for AFP at 15 to 18 weeks. If two repeated blood tests show readings outside the normal range, amniocentesis as well as level 2 ultrasound are strongly recommended. These tests should pick up structural defects that might be due to workplace exposure to chemical substances, radiation, and infections. A few other blood tests are available for specific exposures. New blood assays continue to be developed, and additional prenatal diagnostic testing will become available. Contact a genetic counseling clinic.

Chorionic Villus Sampling (CVS)

Chorionic villus biopsy is the only method for diagnosing genetic disorders that can be performed in the first trimester of pregnancy. The chorion is the membrane that encases the amniotic sac containing the developing fetus and it is derived from the fetal cells. Therefore, the cells from the chorion are genetically identical to the fetal cells. Analysis of chorionic tissue provides much of the same information as amniotic fluid and cells (except for alpha fetoprotein) but much earlier in the pregnancy. CVS can also obtain cells containing enough DNA to do restrictive enzymes techniques now being used to identify markers (known genes that are almost always inherited along with the defective gene which has not yet been found) for a genetic disease such as Huntington's disease or the defective gene itself as in the case of cystic fibrosis.

CVS is a relatively new technique and research has been carried out worldwide to determine its effectiveness and safety. Because there is a high incidence of spontaneous abortions during the first three months, it is hard to detect which miscarriages are due to normal early fetal loss and which to the procedure itself. Recent studies, however, find the rate of miscarriage to be only slightly higher to that from amniocentesis if carried out by physicians well trained in the technique. The prospect of a first trimester diagnostic technique is appealing to most women. One of the main drawbacks of amniocentesis is that it can be done only in the second trimester after the mother feels life and has bonded with her unborn child.

GENETIC TESTING

Genetic testing in the workplace has two facets—genetic monitoring and genetic screening. Genetic monitoring involves examining individuals periodically for environmentally-induced changes in their genetic material. Genetic screening tests individuals for certain inherited traits. The assumption is that either genetic predisposition, exposure to environmental agents, or a combination of both may predispose the worker to occupational diseases. Changes in the egg or sperm could then cause a baby to be born with birth defects.

Genetic monitoring involves collecting blood or body fluids from a group of workers periodically to assess whether genetic damage of the cell has occurred. The procedure focuses on the risk for the exposed group as a whole, because scientists cannot tell which individuals in the group face an increased risk. Ideally, genetic monitoring, if sufficiently improved and used to benefit the worker, could act as an early warning system by indicating that exposures to suspected reproductive or other health hazards are too high or that a previously unsuspected substance is causing harm.

In addition to specific genetic monitoring, there are four other kinds of monitoring activities relevant to detecting reproductive hazards in the workplace:

> *Environmental Monitoring* (EM)—attempts to provide a quantitative estimate of the dose of toxic exposure. It measures concentrations of the agent and not the amount inhaled or ingested by the exposed worker. This is referred to in the technical literature as a measure of "intake."
>
> *Biological Monitoring* (BM)—attempts to measure the actual total "uptake" (intake × absorption) of the toxic agent by the exposed worker. This could occur through several pathways simultaneously (mouth, skin, lungs).
>
> *Health Surveillance* (HS)—periodic medical, physiological, and biochemical examination of exposed workers with the aim of preventing occupationally related diseases.
>
> *Biological Effect Monitoring* (BEM)—the attempt to measure and assess early biological effects of workplace substances whose relationship to harmful reproductive and other health impairments has not yet been firmly established.

Genetic testing is another technological advancement that can benefit or harm pregnant workers depending on how it is used. Once again, businesses might take the "blame the victim" route and use this information to "protect" the more susceptible women by removing them from their workplace into "safer" but lower-paying jobs. In the past, large scale genetic screening programs have been mishandled and misused. Sickle cell screening led to misinformation equating healthy individuals carrying a recessive gene with those actually suffering from sickle cell disease. This resulted in stigmatization and job discrimination, a fate similar to that of many pregnant workers in the past.

PREGNANCY HAZARD HOTLINES

The following centers provide a regular telephone information service about drugs, occupational and environmental exposure during pregnancy, and genetics. Simple questions can usually be answered on the phone, but more complicated ones require a visit with a trained counselor. Most centers provide telephone answers free of charge and have a fee scale for office visits. Check with your medical insurance company or personnel office to see whether genetic diagnostic and counseling clinic services are covered under your policy. Many of the centers only provide information over the phone for in-state residents, but the clinics usually can refer you to sources of information nearer to your home.

1. California Teratogen Registry, University of California, San Diego. California residents only. 1-800-532-3749.
2. Genetics Unit/Teratology Service, University of Colorado Health Sciences Center, Denver, CO. Primarily for residents of the region. (303) 394-8742.
3. Connecticut Pregnancy Exposure Information Service, University of Connecticut Health Center, Farmington, CT. Connecticut residents only. 1-800-325-5391.
4. The Teratology Service, University of Miami School of Medicine, Miami, FL. Answers questions on the phone but is primarily a clinical service. (305) 547-6464.
5. National Birth Defects Center Pregnancy and Environmental Hotline, Kennedy Memorial Hospital, Brighton, MA. Massachusetts residents only. 1-800-322-5014.
6. Teratology Information Network, University of Medicine and Dentistry of New Jersey, School of Osteopathic Medicine, Camden, NJ. Accepts out-of-state calls. (609) 757-7869.
7. Pregnancy Healthline, Pennsylvania Hospital, Philadelphia, PA. Accepts out-of-state calls. (215) 829-KIDS.

8. Department of Reproductive Genetics, Magee-Womens Hospital, Pittsburgh, PA. Primarily serves residents of Pennsylvania and eastern Ohio. (412) 647-4168.

9. Pregnancy Safety Hotline, Western Pennsylvania Hospital, Pittsburgh, PA. Serves Pennsylvania, parts of West Virginia and Ohio. (412) 647-SAFE.

10. Texas State Department of Mental Health and Mental Retardation, Genetic Screening and Counseling Service, Denton, TX. Primarily serves Texas residents and those of bordering states. (817) 383-3561.

11. Texas Teratogen Service, University of Texas Health Science Center at Houston. Will answer out-of-state calls but usually provides a referral. 1-800-835-8360 (Texas only); (713) 792-4592.

12. Pregnancy Riskline, University of Utah Medical Center, Salt Lake City, UT. Primarily serves Utah. 1-800-822-BABY (Utah only); (801) 583-2229.

13. Vermont Teratogen Information Network, University of Vermont, Burlington, VT. Serves Vermont only. 1-800-531-9800.

14. Washington State Poison Control Network and Central Laboratory for Human Embryology, University of Washington, Seattle. Primarily serves Washington residents but will accept calls nationwide. 1-800-732-6985 (Washington state only); (206) 526-2121, (206) 543-3373.

15. Wisconsin Teratogen Project, University of Wisconsin Center for Health Sciences, Madison, WI. Answers some questions on the phone, but not primarily a telephone hotline. Primarily serves Wisconsin, northern Illinois and Michigan's Upper Peninsula. 1-800-362-3020; (608) 263-1991.

16. Teratology Hotline, Birth Defects Center, Children's Hospital of Milwaukee, Milwaukee, WI. Serves primarily the Milwaukee area. (414) 931-4172.

17. Motherrisk, Division of Pharmacology, Hospital for Sick Children, Toronto, Ontario. Primarily a clinical service, individuals calling from outside the area can be counseled by phone. (416) 598-5781.

SUGGESTED READINGS

Barlow, S. M., and Sullivan, F. 1982, *Reproductive Hazards of Industrial Chemicals: An Evaluation of Animal and Human Data*, Academic Press, London.

Chernier, N., 1983, *Reproductive Hazards at Work: Men, Women and the Fertility Gamble*, Canadian Advisory Council.

Hatch, S. L., 1982, "The Psychological Experience of Nursing Mothers Upon Learning of a Toxic Substance in Their Breast Milk," *Psychiatry*, 45.

Kipen, H., and Stellman, J. 1985, *Core Curriculum: Reproductive Hazards in the Workplace*, American Association of Occupational Health Nurses, Atlanta.

Rogan, W., 1986, "Breastfeeding in the Workplace," *Occupational Medicine*, July-September.

Weaver, C., 1986, "Toxics and Male Infertility," *Public Citizen*, 7:2.

PART II.

THE WORKPLACE: HOW SAFE IS SAFE?

Photo © Earl Dotter

Chapter 3

A Comfortable Workplace: What It's All About

> My job as a telephone equipment installer involved pole climbing, ladder handling, and the running of 25 and 50 length cables.
>
> My doctor didn't realize what I was doing, but once he said, "whatever you are doing keep it up, you're in excellent physical condition." The only time he did find out about my job was when I was going into my eighth month and I asked him if he would write me a note saying that I wasn't allowed to climb poles anymore.
>
> —Belinda, telephone equipment installer

While we all may not be as physically fit or as adventurous as Belinda, pregnancy is certainly no longer viewed as an unemployable condition. In 1977, the National Institute for Occupational Safety and Health (NIOSH) issued guidelines on pregnancy and work affirming that a healthy pregnant woman could perform her normal work tasks throughout most, or all, of her pregnancy. These guidelines put to rest many of the outdated views unduly limiting pregnant women's full participation in the workplace.

All workers require safe and healthy workplaces, yet almost all work settings have problems with at least one aspect of the environment—excessive noise, poor lighting, indoor air pollution, poorly designed workspaces and equipment, heavy lifting, extreme temperatures, unsafe and unclean facilities, and most importantly, physical, psychological, and social stress. These problems are not limited to heavy industry or factory work. The secretary who lifts a box of computer paper from a bottom shelf or stands on a shaky chair in order to reach a box of order forms on a top shelf is at risk of a back injury. A receptionist who answers the phone at an automobile body shop is subjected to the same toxic fumes as the painter and welder. Pregnant workers, however, because of physiological changes due to pregnancy, temporarily become even more susceptible to potentially hazardous conditions faced by workers across a wide range of occupational groups.

Canadian researchers studying pregnant workers in 42 occupational groups found excessive rates of miscarriages across all occupational groups for women whose jobs required heavy lifting, other strenuous physical effort, and long hours of work. Exposure to noise, cold, and extended periods of standing also were associated with a higher than expected miscarriage rate, and stress—physical, psychological, and social—was related to adverse affects on pregnancy.

The French National Institute of Health and Medical Research (INSERM) found similar results in a sample of more than 2000 women who had worked more than three months during their pregnancy and gave birth in 1981. The scientists found that women whose jobs involved service work or manual labor had a significantly higher rate of premature births than women with professional or clerical jobs. This difference, however, may be partially due to differences in social class as poorer women are also more likely to have poorer pregnancy outcomes. Assembly line work and jobs where standing, carrying heavy loads, and physical exertion were continuous were related not only to premature delivery but also to lower birth-weight infants. It is not surprising that employers in the workplaces involving the most tiring and arduous tasks for their female employees were the least likely to modify the working conditions. These women were forced to use their sick leave and were more likely to quit work after their second trimester than were women whose employers were willing to modify their working conditions during pregnancy. Unfortunately, it is the pregnant women in the low-paying jobs who need their earnings the most who are forced to make the choice between their reproductive health and their employment.

These issues applicable to many work settings are discussed in this chapter. Chapters 4 and 5 focus on toxic substances and other hazards in specific occupations.

YOUR BODY CHANGES IN PREGNANCY

Your first step in improving your comfort at work is to understand the biological changes in your body during pregnancy and how they affect you. Next find out how the workplace environment interacts with these body changes. For example, if you suffer from neck, back, and eye strain from sitting in front of a video display terminal (VDT) all day, you may lose muscle tone and harm your back, both of which can be made worse over the course of your pregnancy. Last and most important is to improve the fit between your workplace and your pregnancy.

Most pregnant women are aware of the obvious anatomical, physiological, and psychological changes during pregnancy. In early pregnancy, fatigue and nausea often affect comfort and efficiency. In late pregnancy, a protruding abdomen and increased weight and retention of fluid can make walking or changing positions awkward and take some of the pleasure out of even a challenging job. Quick mood shifts and exaggerated emotional reactions may also occur.

We seldom give thought to the more subtle changes, or even the reasons for the more obvious changes. We frequently realize that poor workplace conditions aggravate discomfort but not how proper facilities can alleviate it.

Major physiological changes occur in the circulatory, respiratory, muscle-skeletal, and endocrine systems. The volume of blood increases 30 to 40%. Higher heart rates are normal during pregnancy and they rise more than usual when participating in strenuous activity. You may be particularly prone to this increase if you were anemic, sat around a lot, or were very overweight prior to pregnancy. Furthermore, blood sugar levels are naturally lower during pregnancy and pregnant women tend to burn carbohydrates faster. This combination creates the risk of an abnormally low blood sugar level during periods of heavy activity.

During pregnancy, you may hyperventilate. This means that you move more air through your pulmonary system to extract a greater amount of oxygen than when you are not pregnant. This hyperventilation is due to the elevation of the diaphragm by the uterus and by a 20 to 30% increase in the amount of oxygen your body consumes while pregnant.

You also must adjust to a different level of hormones. The hormones produced by the pregnancy cause ligaments, tendons, and other connective tissues to soften, making you more prone to injury if you carry out tasks that excessively stress or stretch your joints. The expanding uterus and loosened joints put increased stress on your back during pregnancy, and it remains sensitive to injury for several months after delivery.

In the later months of pregnancy, you may fall more easily because the body's center of gravity shifts and throws you slightly off balance. At this time, moreover, the enlarging uterus may press on several organs. It can press on the sciatic nerve, making your legs buckle, or it can press on the large vein carrying blood from the lower parts of your body, causing your legs to swell. Constipation, hemorrhoids, and varicose veins can cause discomfort, and the baby's head deep within the pelvis may press on the base of the bladder causing frequent urination. You can suffer from fatigue, labored or difficult breathing, and insomnia resulting from the combination of the effects of weight gain, increased respiratory requirements, and the discomfort of maneuvering a more bulky and awkward body.

Once you are aware of these body changes, you need to know how the workplace can be modified to ensure you greater comfort. A comfortable workplace depends primarily on good design of work space and equipment and a reduction in physical and social stress. Any modifications that improve conditions for a pregnant worker also makes the workplace healthier and more comfortable for all workers of both sexes and leads to greater efficiency and lower absenteeism. These points are worth emphasizing to co-workers and employers so they do not believe that pregnant workers are receiving preferential treatment at their expense. Another way of adapting work demands to body changes is by contracting for an alternative work schedule. These flexible arrangements can help you to adapt to physiological changes in your body while maintaining your comfort on the job.

If morning sickness is a problem for you, then a schedule starting later in the day can alleviate the tension between your work and your pregnancy. If fatigue is a problem, then a reduced workweek may be an answer. If combining household chores and errands proves difficult, a 4-day/9-hour per day workweek may be your salvation. Your stress might be reduced if you could concentrate on your paid work during these four days. This would give you one weekday to take care of household tasks, shopping, and health care needs and would leave more quality time to spend with your family and friends over the weekend. You could receive all these stress-reducing benefits without loss of income or fringe benefits, and without putting your job in jeopardy by excessive absenteeism or putting your pregnancy in jeopardy by working a schedule that stresses your physiological and psychological capabilities to their limits.

Flexitime—varying arrival and departure time—and the compressed workweek where longer hours are worked on fewer days are the two most common plans. Only about 18% of female wage and salary full-time employees work a nontraditional work schedule by choice. Another option is job sharing. Job sharing consists of one full-time job held by two part-time women. They each work either several hours a day or certain days a week. This has not caught on to any great extent. Only 1% of women workers hold such positions. With a little planning and coordination this option works well. The employer receives reliable service; the job sharers receive fringe benefits, hold down a more interesting job than the usual part-time one, and can spend more time with their children. Obviously, the disadvantage is the loss of income which is necessary for many two income families as well as for women who are the main breadwinners. Some hospitals that suffer from severe nursing shortages have experimented with job sharing and other types of alternative schedules.

Flexitime should not be confused with working swing shift, night shift, or weekends out of necessity. Many women reluctantly work these hours because they need the higher pay, their employers require it, or they cannot afford child care costs and their husbands are home to take care of the children at these times. Flexitime and the compressed workweek are voluntary choices, and with good will on both sides, pregnant workers and employers could work out many more kinds of arrangements. The federal government has also installed flexitime options for most of its own employees. Greater satisfaction leads to a reduction in stress, which is considered to be a reproductive health hazard.

WORK EQUIPMENT, WORK SPACE

Most of us have never heard of ergonomics—the science that attempts to adapt working conditions to suit the worker. Ergonomists, often called human engineers, design adjustments in the work environment to improve safety and efficiency while protecting the health of the worker. In the case of the pregnant worker, ergonomically-sound equipment not only helps protect your body but that of your unborn child as well. Ergonomically-designed work equipment and workspaces, allowing the flexibility to meet individual needs, should be a priority demand for pregnant workers in all types of jobs—executive, white collar, and blue collar.

The equipment women use is usually designed for the "average employee." Your body during pregnancy certainly does not fit that description. In fact, there is no average in a work population consisting of males and females ages 17 to 70. Pregnancy is only one factor in a highly diverse work force by any criteria—size, shape, age, and health.

The human engineer can design flexible equipment that will both increase your comfort and improve the consistency and efficiency of your work. For example, continuous standing is a problem for many pregnant workers, as it can lead to swelling in their legs. Many jobs, especially in the women-dominated service industries, require long periods of standing. Bank tellers often have to remain on their feet all day if their customer windows are high. They would feel a lot better and make fewer errors if they were permitted to alternate their work positions between standing and sitting. Seats, desks, and workbenches of adjustable heights in offices and factories would solve this kind of problem.

Moreover, adjustable tools, tables, desks, benches, and chairs can substantially reduce pressure on joints and muscles that comes from sitting for hours in awkward positions. Women in the garment industry as well as keypunch operators suffer from tenosynovitis, a tendon inflammation resulting from repetitive finger and hand movements in the same position. Sitting over a sewing machine or at a word processor for several consecutive hours also causes muscle strain in the shoulders and back.

Moveable lights, well laid out work space, footrests, and work holders make it possible for you to adjust the equipment to fit your individual needs. This is of particular value to you as your body changes substantially over the 9-month period. Such flexibility allows your body to be less stressed before you become pregnant and to return to its non-pregnant state more rapidly.

Good seating is another priority for the pregnant worker. The backrest of the chair should be low enough to support the lower back and pelvic area. Your seat should be cushioned, flat, and wide enough to allow you to be seated with both legs in a supported position. A footstool is an added bonus in that it takes pressure off the legs and improves circulation.

Workplace design also includes space, light, noise, and air quality factors. If not satisfactory, these components of the work environment can pose harm to you and your unborn children by causing injuries and accidents and through physical stress.

Cleaning up the workplace is the ultimate goal we strive for, but sometimes this is not immediately feasible. Wearing well-fitting personal protective equipment (PPE) such as respirators, dust masks, gloves, and ear protectors may be a necessary, though disagreeable, interim measure. Women, though, seldom have well-fitting protective equipment. This is a prime area of concern. If PPE's are not the correct size, they do not provide adequate protection when worn and if they are uncomfortable, they are not worn.

At present the lack of enough sizes in protective equipment is a problem for women. Some of the personal protective devices are too big as they are often manufactured only in men's sizes. Face masks and shields leave gaps so that dust and fumes enter. Hand tools are often too heavy to handle and hold and too clumsy and difficult to grasp. This leads to undue strain and accidents.

The problem of inadequate PPE's is even greater for the pregnant worker. She may be more vulnerable to hazards than her non-pregnant co-worker and more likely to feel more uncomfortable due to her body's changes. Personal protective equipment that fits in the third month of pregnancy may not by the sixth. As anyone can see, women are not proportionately smaller than men in all body dimensions. In fact, women are proportionately larger than men in three areas: chest depth, hip circumference, and back curvature at the hip. Yet at every height-weight combination, women have significantly smaller shoulders. A pair of coveralls could thus be too small across the hips and too large across the shoulders. As the body changes during pregnancy the problem of adequate personal protective clothing worsens.

Equipment also needs to be redesigned to fit women's lifting capabilities. A lifting task exerts 15% more stress on a non-pregnant woman's back than on a male of the same size or strength because women are built differently from men. This stress is even greater for the pregnant woman, especially in the last trimester when women are more prone to back injuries. Equipment geared to women's lifting capabilities would also be better suited for many men, as continual lifting accounts for a very large number of severe and disabling back injuries—one of the most common causes of industrial absenteeism. Poorly designed equipment is frequently responsible for these injuries rather than any inherent body weaknesses. Ask your employer to inquire about new developments in equipment design from the companies that supply their equipment.

Before you approach management for improvements, it is a good idea to do some homework and figure out ways of presenting your suggestions in terms of long-run savings due to increased efficiency and decreased injury and absenteeism for all workers, not just pregnant ones. Ease of maintenance and inexpensiveness of repairs is a prime consideration for the employer who is considering making improvements in the workplace that would make you more comfortable.

STRESS: THE WOMAN CAUGHT IN THE MIDDLE

Stress is a well-recognized work hazard, particularly in jobs characterized by high demand and low control—just the type of jobs that are overwhelmingly occupied by women. Even when men fill the same positions, they are given greater scope for decision making than are women.

Until recently, scientists have not been interested in studying reproductive hazards of the workplace. Too often we have to rely on some animal studies, a few isolated studies on small groups of humans, and anecdotal reports from pregnant workers. We know that many unpleasant work conditions such as shift work, poor ventilation, or being unfairly reprimanded by our bosses makes us feel uncomfortable and sometimes ill, but we do not always know whether our fetuses will be harmed by this stress. Unfortunately, scientists in many cases can only indicate the possibility of harm but cannot provide any firm conclusions (see Chapter 6 for more information about evaluating risks and hazards).

If you are in a job that demands a lot physically and emotionally, one in which the work pace is quick, then you probably experience stress. If in addition, you cannot control the pace, the time to take breaks, or the planning of your work schedule, you may over a lifetime increase your risk for a wide variety of health problems. These are some primarily women's occupations which have high stress levels characterized by high demand and low control:

Highest Stress Level

Waitress, bank teller, cashier, nurse's aide, assembler, file clerk, receptionist, and sewing machine, telephone, office machine, and keypunch operators

High Stress Level

Cooks, salesclerks, typists, secretaries, library clerks

These are causes of stress:

1. No control over speed of work
2. No control over variety
3. Repetition of monotonous work
4. Not able to leave workstation
5. No control over decision making
6. Insufficient time to complete assigned work
7. Insufficient opportunity to obtain a promotion

In addition to the on-the-job stresses, the pregnant worker often finds herself in a particularly stressful situation trying to combine the roles of worker, mother or mother-to-be, and homemaker. Supermom is not a figment of journalists' imaginations but an accurate portrayal of the diverse roles women play in their daily lives. The term is particularly apt for the pregnant worker who has other children at home.

Physicians see a substantial number of patients suffering from stress and now consider it a factor in a vast majority of physical and mental ailments. Figure 3.1 shows that a wide variety of health problems may be stress related. Birth defects are included as possibly having a stress component. This possible link between stress experienced by the pregnant woman and a baby being born with a birth defect is a worrisome one. Some writers have even suggested that stress caused by a variety of factors including interpersonal tensions may affect the development of the fetus leading to neurological difficulties, physical defects, slow development, and behavior disturbances in the child. These connections, however, are far from certain.

Work-related stress can arise from physical conditions of the workplace or from interpersonal sources such as our relationships to our own job tasks. Our jobs can be monotonous and monitored, our co-workers unpleasant and uncooperative, our superiors unfair and unsympathetic, and management's policies can be inequitable and inflexible.

Despite the evidence that sources of stress stem in the main from the physical and interpersonal structure of the workplace, many doctors, books, and courses advise pregnant women to handle stress by individual techniques such as relaxation exercises, yoga, and meditation. These are helpful. But why put the burden on the individual to counteract the stress rather than reduce it at the source?

Table 3.1 lists some of the main workplace conditions we should be concerned about. The pregnant worker can learn about the effect of physical and interpersonal workplace stress and then use this knowledge to try to change working conditions. (Chapter 7 provides suggestions for taking action.)

FIGURE 3.1. Health Problems Thought to Be Associated with Stress at Work

accidents	birth defects	indigestion
alcoholism	colitis	infections
anxiety	depression	insomnia
arthritis	fatigue	neck pain
asthma	headaches	skin rashes
backaches	heart attacks	strokes
battering	high blood pressure	ulcers

Material drawn from Health and Safety Guidelines for Health Care Professionals, University of Connecticut Health Center Professional Employees Association.

TABLE 3.1. Some Major Workplace Conditions That Cause Physical and Social Stress to Which a Pregnant Worker Is More Vulnerable

Noise and vibration	Lack of adequate sanitary and safety precautions
Poor lighting and glare	Shift work
Inadequate ventilation	Friction with boss and co-workers
Extreme heat and humidity	Child-care difficulties
Lifting heavy weights	

Excessive Noise and Vibration: Can the Fetus Be Harmed?

Exposure to a lifelong work environment of loud noise has damaged hearing and even caused deafness among workers. Assembly line workers, airline attendants, garment and textile workers, and construction workers among others are exposed to a great deal of continuous noise and vibrations as part of their jobs. The revving of airplane engines, continuous clanking of the assembly line, whirring of sewing machines, and the roar of construction machinery take a health toll on workers in these fields. One pregnant assembly line worker who made automobile seats complained that all she could hear was the noise of the constantly running machines and attributed her stopped-up ears to this continuous noise. Noise affects more than hearing and can cause more than headaches. It can make your blood pressure rise and long-term exposure may be a contributing cause of heart disease. By blocking out warning signals, noise also constitutes a safety hazard.

But can a high level of noise and vibrations also injure the fetus and affect the pregnancy? We know from studies and personal anecdotes that the fetus responds to the noise and vibrations in the mother's environment. One study found easily distinguishable fetal movements when an automobile horn honked a few feet from the pregnant woman. In several cases, mothers reported fetal movements immediately after sound stimulations. A pregnant miner who worked in an area where there was noise from sledgehammers and cars carrying coal reported that her fetus was more active when she worked in the mine and quieted down after she left work. Whether this receptivity to noise external to the womb causes it harm is not definitively known.

Results of animal studies do raise suspicions as to the safety of occupational noise and vibration regarding reproduction. Some of the pregnancy effects seen in animals exposed to noise-induced stress are spontaneous abortion, premature delivery, and toxemia. One hypothesis is that exposure to excessive amounts of noise and vibration disturbs the circulation of the mother's blood in the uterus. The combination of information gathered from animal studies and human adults suggests that it would be wise to avoid undue noise and continuous vibrations while pregnant until conclusive evidence is gathered regarding the fetus's vulnerability.

Machines can be designed to minimize noise and vibration, and proper maintenance can help reduce the noise of existing machinery. Noisy processes should be isolated in one section of the workplace using noise absorbing material in floors, partitions, walls, drapes, and ceilings. By these measures and the provision of frequent rest periods and job rotations, your employer can decrease your exposure.

Inadequate Ventilation: What Are We Breathing?

Unless a building is properly ventilated, pollutants can build up to levels that are unhealthy and possibly dangerous to us and our unborn children. This is the situation found in many workplaces where indoor air pollution is a severe problem. Construction of plants, factories, hospitals, schools, and office buildings incorporating energy saving measures with windows that do not open have led to this grave situation. Cigarette smoking, carbon monoxide emissions from loading docks and garages, fumes from office and factory machinery, and the increased number of chemicals used in work processes, furniture, and equipment make the matter worse. Smoking by a pregnant woman is linked to a higher risk of her having a low birth-weight baby. Passive smoking (the cigarette smoke in the air breathed in by nonsmokers) is now connected to an increased risk of developing lung cancer or emphysema. Researchers, however, have not yet determined the increased risk of passive smoking, if any, to the pregnant worker. OSHA is looking at the issue of indoor air pollution, particularly with regard to passive smoking, and is deciding whether to issue an indoor air quality standard.

Exposure to several toxic substances such as carbon monoxide, lead, benzene, and hydrogen cyanide can interfere with the oxygen-carrying capacity of the blood. In addition, it can cause cancer, respiratory problems, nervous disorders, and sterility. We should limit our physical effort in workplaces where these chemicals are in the air. Inhaling air containing these chemicals may cause us to suffer only minor symptoms, but fail to alert us to the more serious fetal effects.

> There was solder smoke. They said that was not really good for you by itself and even worse if you were pregnant.
>
> —Marcy, solderer and wirer

> I do feel that my baby was at risk because of the smoke in the unit and the lack of ventilation. There are studies now that passive smoke is more dangerous than it was previously thought to be. There are problems with low birth-weight babies in places where there is smoke, so I do feel that the baby is at risk. I really don't know yet because I haven't had the baby.
>
> —Adie, psychiatric nurse

The key to cleaner air in your workplace is the type and maintenance of the ventilation system installed in the building. These ventilation systems should be designed to supply and circulate fresh air, and to filter or eliminate contaminated air. The circulation of fresh air by itself is not necessarily a satisfactory solution, as the outdoor air may also contain pollutants. Unless the system is correctly designed and in very good working order it will not be able to eliminate even moderate amounts of toxics from either fresh or recycled air.

The blower that moves the air, the ducts that deliver the air, and the vents that distribute the air are the basic components of the ventilation system. Often the system fails to adequately supply fresh, clean air. The blower may be underpowered for the amount of space it is supposed to service, or there may be too few ducts and vents, or they are dirty or blocked.

Following are a ventilation checklist and list of references to find out more about indoor air pollution. Check your workplace. You probably can come up with some relatively inexpensive and easy-to-implement suggestions that can be used as interim measures until the basic underlying conditions can be improved. Some of these may involve more frequent maintenance checks of the ventilating system, changing from a toxic cleaning agent to a non-toxic one, or moving boxes or furniture located in front of the ducts.

Ventilation Check List

1. See if your workplace has a ventilation system. You can do this by looking for ducts and vents.
2. Is your ventilation system on 24 hours a day if large duplicating and printing jobs are done at night? These machines can produce a high volume of pollutants. Hold a tissue near the vent after 5 p.m. If it moves, then air is circulating.
3. Does the ventilation system go on and off during the day? Some systems are on a time cycle regulated by a computer. This type of system gives inadequate amounts of clean air if pollutants are generated continually. Use the tissue test described above several times a day for a few days at different hours.
4. Does each room have a supply vent and an exhaust vent? Again use the tissue test and see whether air is entering.
5. Are the exhaust and supply vents right next to each other or blocked? If the two types of vents are too close to each other, the clean air gets sucked out of the room before it has a chance to circulate. If the vents are blocked by walls, partitions, boxes or files, the air flow will be obstructed and pollutants will not be eliminated efficiently.
6. Are there "dead spaces" in your workspace where no air is being replaced? Light a match and if the smoke does not move you can be pretty sure that the pollutants do not move either and just build up.
7. Do work areas with copying and printing machines have adequate air supply and exhaust? For some machines extra vents near the source of the fumes are needed.
8. Can workers control their ventilation by turning the blower or fan supplying the air up or down? Check with your building maintenance office to see whether this is possible.

9. Is there a smoke detector in your ventilation system? For early detection of a fire, it should be located in the exhaust vent.
10. Are the temperature and humidity adequate? Humidity makes a cold room feel colder and a hot room feel hotter. If the air is too dry, you may become more susceptible to colds and infections.

Local health departments usually require minimum ventilation standards for workplaces. These vary from state to state. If the indoor air pollution at work is causing a problem, check with the local, county, or state health department listed in the telephone directory to determine if the employer is violating the health code.

Several consumer groups have issued pamphlets on indoor air quality: "Indoor Air Quality . . . A Number One Priority," from the Consumer Federation of America, 1424 16th Street, N.W., Washington, DC 20036; "Air Pollution in Your Home?," "Home Indoor Air Quality Checklist," "Indoor Air Pollution in the Office," and "Office Indoor Air Quality Checklist," from the American Lung Association, available from your local Lung Association; and "Indoor Air Quality Notes," by Thad Godish, PhD, a series of 6-page fact sheets on important indoor pollution problems (send stamped, self-addressed envelope to: Natural Resources Department, Ball State University, Muncie, IN 47304).

Extreme Heat and Humidity: It's a Long, Hot Summer

Workers in bakeries, canneries, laundries, and garment and textile factories frequently work in hot and uncomfortable environments. You may feel ill or dizzy when exposed to such extreme heat and humidity at work.

> For the first two weeks of my pregnancy it was 140 degrees up here and I was getting dizzy. I couldn't figure out why because I'm so used to the heat. So I knew I had to be pregnant.
>
> —Betsy, finisher and presser
> in dry cleaning establishment

If these conditions are severe or prolonged, not only can you suffer increased fatigue and discomfort, but the more serious conditions of heat exhaustion and heat stroke as well.

Pregnant women are more sensitive to high temperatures because their bodies have to rid themselves of heat produced by their own increased metabolic rate in addition to displacing the fetal body heat which is about one degree higher than their own.

If you engage in vigorous activity while pregnant for more than 15 to 20 minutes at a time your body temperature may increase. This usually does not cause problems. But if this occurs frequently, particularly when the weather is hot and humid, or when the workplace is hot and stuffy, your unborn child could be harmed. The fetus does not have the capability to cool itself, and your overheated body will not be able to accomplish its normal task.

Dehydration is an additional hazard of jobs requiring strenuous activity in a hot environment, as is the loss of body salt. You can prevent this if you drink large quantities of water or other fluids and limit or space out your activities. Less body heat is produced during intermittent high periods of activity than during steady heavy work. Dehydration can interfere with the amount of fluid and nutrients the fetus receives and can trigger early labor. Increased demands on the circulatory system can impair alertness, mental functioning and physical capabilities. Pregnant women feel particularly uncomfortable when there is an extreme temperature change and should avoid work situations involving drastic temperature shifts such as those that occur when entering and leaving cold storage rooms or hot furnace areas. Try to accommodate your pregnancy by minimizing the amount of work you have to perform in such areas or by taking more frequent breaks.

Poor Lighting and Glare: Vision Problems on the Rise

Lighting problems are not thought to affect the reproductive functions of workers or the health of the fetus directly. Indirectly, improper lighting can cause a great deal of discomfort for the pregnant worker as light affects comfort, safety, efficiency, and mood as well as the ability to see. Both extremes, too little light as well as glare from light, cause difficulties. Eyestrain from VDT (Video Display Terminal) work, a universal complaint, has been documented in many studies. In fact, a majority of women in a survey taken by 9 to 5, the National Association of Working Women, responded that lighting was the most important physical aspect of their workplace. Inadequate or improper lighting is a widespread problem and is usually not an isolated factor, but an indication of generally poor physical working conditions.

The precise amount of light we need depends on our eyesight, age, and the task we are doing. The finer and more detailed the work, the higher the level of light needed. For example, performing intricate tasks under a microscope in producing microchips and circuits in the electronics industry, or performing close, precise bench work tasks may cause you to hunch over into a position that puts pressure on your neck, particularly in the later months of pregnancy when the center of mass of your body has shifted forward. This is made worse when either bifocals are worn or there is inadequate lighting causing eyestrain which then spreads to the head and neck compounding pain and discomfort. Whenever vision correction is needed for this type of work, intermediate or near vision glasses rather than bifocals may be better.

The effect from illumination glare requires further research. Glare is caused by light shining directly into the eyes or bouncing off a surface and reflecting into the eyes. Glare can be reduced by shielding the lights with plastic or devices called diffusers that spread out the light. These shields are relatively inexpensive. Glare can be further reduced by eliminating shiny, reflective surfaces on furniture and equipment. Reducing glare from the VDT screen has led to the manufacture of auxiliary devices, and glare-free screens have become a selling point in the highly competitive VDT market. (Chapter 5 will provide further details about the hazards of VDT work and what to do about them.)

In order to conserve energy or to save money, employers frequently reduce lighting levels without checking whether dimmer lighting meets the recommendations of the Illuminating Engineering Society incorporated in many building codes. Poor maintenance is often responsible for dim lights. Light bulbs that are not dusted can reduce the light level by approximately 5% a year. If this is a problem in your workplace, suggest to management that they clean fixtures, replace defective equipment, and filter air to reduce indoor pollution. These relatively simple and inexpensive measures may correct the condition completely, or at least will increase the brightness of your surroundings.

Lifting or Pushing Heavy Weights: Our Vulnerable Backs

A wide variety of traditional and nontraditional jobs held by women, such as nursing; home, office, and hospital cleaning; and construction, laundry, and industrial work require a great deal of heavy physical labor. Lifting or pushing heavy weights is one aspect of work that can lead to your suffering a miscarriage. It puts an extra strain on the pregnant worker.

For instance, Darcy, a hotel maid, sometimes had to lift cases of Perrier water to restock the refrigerators and had to push a heavy metal cart loaded with sheets and towels weighing about 20 to 30 pounds. She attributed her backaches to the strains of these efforts. Jobs in assembly plants also can be extremely strenuous.

> I buffed car seats, trimmed them and threw them on the line. We lifted up heavy frames, 16 pound frames, throwing them on the line, turning them over and turning them over again. I was standing the whole 8 hours bending over, lifting up the car seats, frames, and cushions.
>
> —Carla, automobile seat finisher on assembly line

While both these workers may be overexerting themselves, pregnant women in general can continue to perform familiar occupational tasks requiring a good deal of strenuous activity. This is frequently less stressful than being switched to a new but physically less demanding position which can be more trying emotionally. With some modification, this holds true for jobs that require lifting and pushing heavy weights as well. The general rule of thumb is if the woman can handle the load easily when not pregnant, she probably will not be unduly stressed during her pregnancy except for the last few months. Then it probably is advisable to reduce the maximum load lifted by 20 to 25%. State regulatory bodies have generally discarded restrictions on the amount of weight to be lifted by women whether or not they are pregnant. That limit depends on the physical strength of the woman and how she feels during pregnancy. The strain from pushing or lifting heavy weights varies from woman to woman and can differ from one pregnancy to another.

Proper lifting techniques will minimize injuries to your body or your pregnancy. Avoid lifting in front of your body during the last trimester as it places a burden on the lower (lumbar) region of the spine already under stress by your weight gain, expansion of the uterus, and loosened joints. You should be aware that two adrenal hormones, epinephrine and norepinephrine, increase during strenuous activity. Epinephrine quiets uterine muscles but norepinephrine increases the muscular activity, possibly inducing premature labor. If mechanical lifting aids are not available, you should either divide your burden into smaller units or ask for help.

Lack of Adequate Sanitary and Safety Precautions: Disease and Danger

Pregnant laundry workers, health care personnel, lab technicians, airline flight attendants, and elementary school teachers often come into contact with individuals who have infections or material that is contaminated. We seldom think of flight attendants as being in danger of catching infectious diseases, but Nora, a pregnant flight attendant who has worked in her job for thirteen years, recalls passengers collapsing from hepatitis and children taken off the plane with chicken pox. She says that she finds herself looking at passengers to see if they look healthy and if they do not, trying to guess what illnesses they might have.

Even office workers can face these kinds of hazards. Pamela worked as a general office worker in a residential treatment facility while she was pregnant. Part of her job was to sort clothing that was donated. She was worried that some of the clothing was contaminated because her supervisor told her to throw the dirty clothing out rather than bring it to the patients in the cottages.

Although the pregnant woman does not seem to be more susceptible to infection, once she does become ill, the infection may be more severe and require a longer recuperation period. Certain diseases such as rubella (German measles) can cross the placenta and harm the fetus. So can some drugs given to the mother to combat an illness. These can result in miscarriage, fetal infection, abnormality, or even fetal death.

The following prescription drugs should be avoided during pregnancy if at all possible: dilantin, cortisol, androgens, chemotherapy (anti-folàte agents, alkylating agents), diethylstilbestrol, tetracycline, opiates, and benzodiaepines. Also take extra precautions if your job involves manufacturing or administering any of these drugs. If you are in doubt about taking or administrating a specific drug, call one of the Pregnancy Hazard Hotlines listed at the end of Chapter 2.

Another hazard is improperly treated garbage and disposable waste in hospitals, manufacturing plants, and office buildings. For example, a chemical that should be disposed of in a leakproof container is spilled into a sink instead. In the sink drain it combines with the remains of another chemical that should not have been poured in the sink either. The newly combined chemical then bubbles up into a sink in a different room subjecting other workers to a hazardous substance of which they are totally unaware. Workplaces themselves may be dirty—spills are not immediately wiped up, cockroaches and bugs cohabit, clutter accumulates in the halls and workstations, floors and stairs are slippery, and machinery is not maintained properly. Basic safety measures may also not be followed.

> When you were walking from break, a cushion or frame could fall off and hit you. They would take you to a hospital if you were to cut off a finger or something. But if you get hit with a cushion, they'll let you sit down in the office for 5 minutes or so and put an ice bag on it.
>
> —Carla, automobile seat finisher on assembly line

You are especially vulnerable to existing safety hazards during the last trimester when your body is harder to maneuver and you have little agility.

If inadequate sanitary and safety precautions are typical of your workplace, you can keep records of the specific incidences, documenting any accidents, injuries, and near misses. Summarize some of this information for your co-workers and try to form a committee to seek improvements from your supervisor and employer. Speak to your union or employer and request them to form a health and safety committee which would inspect the workplace on a regular basis and recommend modifications and a timetable for their implementation. The CUNY Center for Occupational and Environmental Health, 425 E. 25th Street, New York, NY 10010, (212) 481-4361 is a good resource, particularly for training factsheets, and for a computer program that helps employers and local unions track injuries in order to find out what may be causing them. See Chapters 4 and 5 for more detailed information about toxic reproductive hazards in specific jobs.

SOCIAL STRESS

In addition to physical stress, social stress poses problems for pregnant workers. This usually stems from a clash between your own economic and social needs and those of management or society. As a result, you can face social stress from several different aspects of your job. Inflexible work rules and shift work compound the normal fatigue of pregnancy. Friction with your bosses and co-workers can make your work life miserable, while unclear guidelines and child care difficulties can leave you continually worried.

Inflexible Work Rules: Give Us a Break!

You may find that inflexible work rules create more of a hardship than the difficulty of your work, as few jobs demand continuous physical activity; rather they require a large amount of exertion only part of the time. As a pregnant worker, you need additional flexibility and you feel the restraint of rigid work rules even more. You need more frequent rest breaks where you can walk around, change positions, stretch, eat a nutritious snack, or use the toilet, rather than a reduction in your work load. Access to restroom facilities when you need them is extremely important because the uterus exerts pressure on the bladder, especially in the last few months. Increased frequency of breaks and flexibility in the timing of breaks allows you to continue performing your job efficiently.

> I was only allowed to use the lavatory on those planned recess periods, or at lunchtime, which was inadequate for my needs. Fortunately, the children's bathroom with the low toilets was just across the hall. I was always running across little kindergartners while I was in there. It was against the rules, but it was really necessary.
>
> –Kathy, elementary school teacher in a private school

Sometimes even when the work rules are more flexible, facilities are inadequate. For example, either there is no adequate lounge where you can rest or put your feet up, or no cafeteria where you can obtain nutritious food.

> We'd have nowhere to go during break. All we had was a bathroom which has two toilets. Sometimes when I was really tired, I'd go in there and sit on a toilet and other times I'd sit on a window sill. If anything, my breaks decreased because my boss wasn't very happy when she found out that I was pregnant.
>
> –Pamela, office worker at a residential treatment facility

Shift Work: Body Rhythms Out of Cycle

In 1980, 18% of employed women reported working on rotating shifts. Medical evidence indicates that such schedules pose psychological and physical hardship because our circadian rhythms (body functions that vary systematically throughout the day, such as temperature and physical energy) often become out of sync with the rest of our activities. This affects digestion, the immune system, sleep, alertness, motor reflexes, motivation and powers of concentration. It is thought to be one reason why the number of work-related accidents reaches its peak between 4 and 6 a.m.

Shift workers on the average smoke more heavily, are more often obese, eat less nutritional foods, participate in fewer leisure activities, and are less involved in social networks. Shift workers also have higher cholesterol and serum triglyceride levels–risk factors for cardiovascular disease. Women who work rotary shifts show high levels of job stress and emotional problems and more frequently use sleeping pills, tranquilizers, and alcohol than their counterparts who work fixed shifts. Pregnancy causes physiological changes, shift work upsets natural body rhythms, and the combination can upset the body's functioning. Therefore, shift workers may be more likely to suffer ill effects from their variable schedules after they become pregnant than they did before.

Excessive and chronic stress should be avoided by everyone, but especially by the pregnant worker. As was discussed previously, many of the body's reactions to stress, such as a rise in blood pressure, an increased heart rate, digestive disturbances, and greater blood flow to the limbs compound existing stressful, and sometimes dangerous, changes resulting from pregnancy. Recent research indicates that emotional stress constricts the blood supply to the uterus and placenta.

This is not to imply that pregnant workers cannot tolerate moderate, occasional amounts of stress. In fact, some stress is useful in reevaluating pregnancy needs on the job. However, chronic workplace stress can cause the pituitary gland to stimulate quantities of adrenalin more appropriate for emergency situations rather than everyday activities and this is unhealthy.

Reducing stress in the workplace is not always expensive. Sometimes small, well-planned changes in the workplace that cost little or no money can make a big difference.

> I worked evenings from 5:30 p.m. to 9:30 p.m. And at 7:30 we had a 15 minute break. We could have eaten in the cafeteria, but it was impossible to reach the cafeteria and eat in fifteen minutes. If the employer could have given us an extra 5 minutes for our break we might have been able to eat dinner in that time.
>
> –Darcy, hotel housekeeper

Even if Darcy's employer is too stingy to give her the extra 5 minutes, she could ask to start her shift 5 to 10 minutes earlier in return for a slightly longer break that would allow her to eat dinner and obtain adequate nutrients for herself and her fetus.

Working night or swing shifts (usually 3 to 11 p.m.) aggravates fatigue of pregnancy and can reduce appetites to an unhealthfully low level. Even if you are hungry, you may not be able to find any food at all because the cafeteria is closed or it is unsafe to leave the building at night. You lose touch with your family and friends and often cannot visit or attend parties as you need to sleep while others enjoy themselves. You may become more isolated and lose a valuable support network, increasing the stress. What is even worse for you is the rotation of shifts which upsets your body's eating and sleeping cycles.

> We had to work overtime and on Saturday and Sunday if there was a lot of work. It was tough because we didn't have set hours every day. Having to get up at a different hour every day, eat at totally whacked out hours would put a strain on any normal person. You could work 9 to 5 one day, 11 to 7 the next day or 2 to 10 the next day. They could have been more accommodating since there were 5 of us. But I think they had a bad attitude. Our supervisor took it as a personal thing that we all got pregnant at once.
>
> –Maria, lab technician

Many shift work jobs are particularly stressful and fatiguing because they are boring, repetitive, and allow you no control over your work. Fatigue and lack of appetite are more than nuisance problems because they increase the likelihood of your giving birth to a low birth-weight baby who faces a higher risk of illness or death within the first few months of life than a baby who weighed over five pounds at birth.

The only time that a night or swing shift provides benefits is when it also provides additional autonomy. This was the case with Pam who worked the night shift in the sterilization department of a hospital. She was able to switch from a sitting to standing position whenever she felt stiff and could work at her own pace because there was only one other worker on duty at night. During the day many employees and supervisors were on duty making sure that the rules and regulations were being followed.

For most of you, shift work does not offer the freedom that Pam had, and you should try to minimize the strain on your body. If you are on a rotating shift, you can request that you do not rotate daily or weekly but stick to one shift for at least three weeks so your body can stay adjusted for a period of time. On some types of jobs this may not be possible, but even when this schedule could be easily implemented, nothing is likely to be done unless you and your co-workers assert yourselves.

Friction with Supervisors and Co-Workers: Dislikes and Disagreements

Pregnant workers report how upsetting they find disputes with their supervisors about their work and pregnancies. Some researchers think that severe interpersonal tensions may affect the development of the fetus, leading to physical defects, neurological impairment, slow development, and behavior disorders in the child. Luckily, most pregnant workers find that their supervisors are understanding and co-workers become protective of their health. When the pregnancy is resented by others, however, your life can be made miserable. Instead of being supported, you are harassed.

The supervisor/supervisee relationship is a very delicate one even when the worker is not pregnant, and a poor relationship can sour other aspects of her life.

The *Women in Work and Family Life* study being conducted at Bank Street College cited improvement in the boss-employee relationship as being the most important factor in women's jobs that would spill over to improving their family life as well.

> My boss wasn't very happy when she found out I was pregnant. She told me right up front that she hated the idea of my being pregnant and wanted to know what I was going to do about it. I did everything I did when I wasn't pregnant and then some because I had to show her that I could still do my work and function the way I would if I was not pregnant. I guess she wanted to break me or break my spirits so she always had something extra and I never knew what that was until I'd come in.
>
> —Pamela, office worker in a residential treatment center

Co-workers can be as significant as bosses in contributing to the stress level of the workplace. Being in continual contact with people who dislike you is extremely draining. Take the case of Caroline, a telephone operator for a doctors' answering service who was harassed by the other operators who showed no concern for her pregnancy. These women suggested that she leave rather than reduce their smoking.

> The women urge you to quit. They would urge anyone to quit because they've been there for years and it's like the new kid on the block. You know you get the hard time. So they're no help. They don't encourage you. We don't work together the way we should. There's too much tension.
>
> –Caroline, telephone operator for a doctor's answering service

Thus the attitude of bosses and co-workers can make or break a job, especially if you are already coping with nausea, fatigue, swollen ankles or backaches.

> I didn't feel that other staff members were supportive during my pregnancy. In other places, people seemed to help pick up and swap off duties when someone wasn't feeling well and I didn't find that the case here. I think there was a difference between the male and the female attitudes. The males seemed to be very protective and wouldn't let me get involved in anything that might be detrimental to my health. They wouldn't let me lift, they wouldn't even let me sit one-to-one with a patient who was having any kind of difficulty at all. The women felt that I could do anything that anyone else could do.
>
> –Adie, psychiatric nurse

Female co-workers may feel hostile or angry toward pregnant colleagues for many reasons: because they fear that they will create more work for them or perhaps they are jealous of women who combine motherhood and work when they felt they had to choose one or the other. As more and more women work outside the home during their pregnancies, skeptics will learn that pregnancy is not a barrier to effective job performance. Pregnant workers may even serve as role models for other women who were uneasy about attempting the joint venture of paid employment and pregnancy.

If you pull your own weight and do not put a burden on others to do more than their fair share, you will minimize discord. If you do need special arrangements during part of your pregnancy, explain to your co-workers and supervisors what the problem is, for how long you will need these arrangements, how they will or will not impinge on the other workers, and how you might compensate for this special treatment. If the work is being completed and no one else is being imposed upon, formal or informal agreements usually can be worked out.

Child Care Difficulties: Who's Going to Watch My Baby?

Before we even give birth to our babies, we think about problems we face when we will return to work or whether we should return at all. Regardless of the reasons for our working – self-fulfillment, economic necessity, or both – being separated from our newborns is a wrench and obtaining child care a worry.

> My boss has always been real skeptical of my getting pregnant and leaving. She would ask me if we could afford to live and not work. I kept answering that we couldn't afford for me not to work, but we found out that we could. And when you start thinking in terms of leaving a baby with someone, it just got real important to me to stay home. So, I did.
>
> – Melissa, employee service administrator at a bank

> I brought up to the plant manager that the factory workers get three months after their initial six weeks after their babies are born and we who work in the office only get the initial six weeks. We were saying how tough it is to find a babysitter at 6 weeks. So his answer was, "O.K., what we're going to do is find a day care center in the area that will take them at 6 weeks old." I would have loved to have six months to a year with

her, if I could have. If they had a day care center at the plant, I would like that a lot.

—Jean, personnel office worker
in automobile plant

Many of you want the option of staying home with your new baby for several months without losing your jobs or seniority. If you decide to return to work, some of you would like to have a company-provided child care facility on the premises or nearby. Often several smaller companies in one geographical area can jointly operate one facility for their employees, sharing space, staff, and expenses. You could then visit your baby during lunch hour and continue nursing if you wished.

Going back to work shortly after giving birth and continuing to work while raising young children is a constant juggling act, and in times of crisis, we drop one of the balls. Workers and employers find that conflicts between jobs and pressures from family duties are becoming increasingly troublesome. Finally these conflicts are reaching the surface and being openly discussed. In the beginning, because they were seen only as women's problems, they were largely swept under the rug. Now they affect male workers as well because most of these family men have working wives with their own employment obligations.

Mothers can no longer be counted on to stay home with a sick child if both incomes are needed. Fathers then have to stay home, arrive at work late, or leave early to fulfill their family obligations. Often mothers and fathers have to alternate taking days off when children are sick so that neither loses his or her job. Single mothers are in an even worse bind if they do not have close family or friends who are willing to come to their rescue in an emergency. These strains have led to high absenteeism and losses in productivity, forcing business to emerge from its cocoon and grapple with solutions. Schools also do not recognize the new American family, or the plight of the single mother, and schedule early closings and special days off for teacher training without thinking about who will care for the children.

While social and physical stress can impinge on the health of pregnant workers in any occupation, certain kinds of jobs pose special hazards. These occupations will be covered in the next two chapters.

SUGGESTED READINGS

Clark, N., Cutter, T. and McCrane, J. A. 1984, *Ventilation: A Practical Guide*, Center for Occupational Hazards, New York.

Cooper, C., Cooper, R. and Eaker, L. 1988, *Living with Stress*, Penguin, London.

Factsheet: Health Problems from Lifting and Carrying (includes exercises for low back pain), from National Union of Hospital and Health Care Employees, RWDSU AFL-CIO, 330 W. 42 St., New York, NY 10036.

Mackay, C. and Bishop, C. 1984, "Occupational Health of Women at Work," *Ergonomics*, 27.

Tasto, D.C., Colligan, M., Skjei, E. and Possy, S. 1978, *Health Consequences of Shiftwork* (NIOSH Publication No. 78-154), National Institute for Occupational Safety and Health, Cincinnati.

U.S. Department of Health, Education, and Welfare, 1977, National Institute for Occupational Safety and Health Research Report, *Guidelines on Pregnancy and Work*.

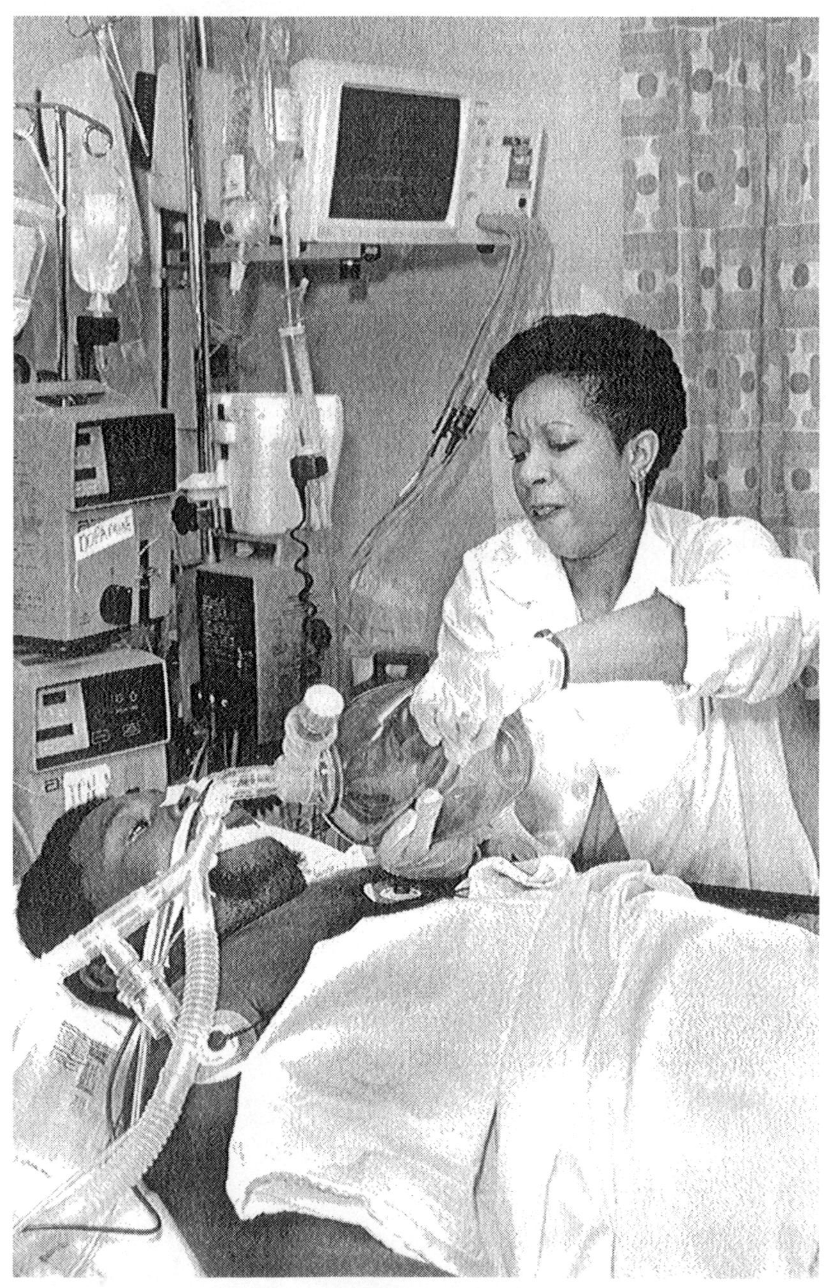

Photo © Earl Dotter

Chapter 4

Women in Health Care, Industry, the Service Sector, and Agriculture

Is My Baby All Right? is the title of a book popular in the 1970s. This is the question that we ask ourselves repeatedly when we are pregnant. We are increasingly asking whether our workplace is conducive to having a healthy baby, especially after we have seen stories about clusters of miscarriages and birth defects among women in various industries. Women who work in health care, industry, agriculture, and the service sector, are often at risk of reproductive harm. They come into contact with many different substances and no publicly available toxicity information exists for more than 70% of the 59,000 plus chemicals on the *Registry of Toxic Effects of Chemical Substances* compiled by NIOSH. Furthermore, OSHA regulates fewer than 600 of the toxic chemicals. Therefore, pregnant workers for the most part have no way of knowing the health implications of the chemicals they work with, let alone the reproductive implications. Too often, only after a group of workers have become sick has the toxicity of the substance become known.

While scientists do not have answers, they have suspicions about many substances and agents, based on animal and bacterial studies as well as the more meager human evidence. The Congressional Office of Technology Assessment (OTA) reviewed the research on the agents and substances in Table 4.1 that showed some indication of reproductive harm. The findings regarding human reproductive damage were inconclusive for almost all the agents and substances. There is general agreement on lead, ethylene oxide (EtO) used in sterilizers, fumigants and industrial processes, dibromochloropropane (DBCP) used in certain pesticides before it was effectively banned, and ionizing radiation being definite reproductive workplace hazards, though there is also a growing consensus about organic solvents and heavy metals. (At the end of Chapter 6, which talks about risks and hazards, there is a list of substances, their types of possible reproductive effects, and whether the data come from human or animal studies.) Infections such as the HIV virus, rubella and cytomegalovirus can also harm the fetus.

The problem pregnant workers face continually is the uncertainty of the information and what to do in the face of such uncertainty. You must make judgments and decisions based on these judgments and avoid the two extreme positions. One extreme is to become very frightened because you believe that all chemicals will eventually be shown to have some harmful human reproductive effect. This is analogous to the position people take that "everything causes cancer," which is not true. The other extreme is to equate the finding that there is no evidence of harm with the belief that there is no harm. This is wrong. There are many reasons having to do with the research study design that might lead to a conclusion that a substance does not appear to be harmful when it really is, particularly when the exposure is low and takes years to show damaging effects. (Chapter 6 goes into this in greater detail.) An understanding of the information needed to judge risks and hazards is crucial in making an informed decision about the potential reproductive harm of your workplace and how you can protect yourself. If you think you are exposed to any of the substances or agents investigated by the OTA as possible reproductive hazards and if you are pregnant or considering getting pregnant, ask your employer for the latest information under your state or federal Right-to-Know Law, check with your obstetrician, call the nearest university department of occupational or environmental medicine, consult your union, or read the evidence in the *Reproductive Health Hazards in the Workplace* report cited at the end of Table 4.1. See Table 4.2 for a list of known and suspected reproductive hazards in various occupations.

If the evidence is *very* weak as to whether a substance is a reproductive hazard, do not worry unduly about the exposure. Constant worry can cause stress, which is known to be harmful to your health. This may be potentially more damaging than the slight possibility of harm due to your exposure to a given agent. You want to minimize your risks as much as possible, but remember that most women do give birth to healthy children. The next sections deal with specific hazards in the health care field, service sector, and some industrial jobs.

TABLE 4.1. Agents and Substances Reviewed for Reproductive Health Effects by the Congressional Office of Technology Assessment

Chemicals:	Physical Factors:
Agricultural chemicals:	Ionizing radiation:
carbaryl	x-rays
dibromochloropropane (DBCP)	gamma rays
DDT	Non-Ionizing radiation:
kepone (chlordecone)	ultraviolet radiation
2,4,5-t, dioxin (TCDD), and agent orange 2,4-d	visible light
Polyhalogenated biphenyls:	infrared radiation
polybrominated biphenyls (PBBs)	radiofrequency/microwave
polychlorinated biphenyls (PCBs)	laser
Organic solvents:	ultrasound
carbon disulfide	video display terminals
styrene	magnetic field
benzene	Hyperbaric/hypobaric environments:
carbon tetrachloride	Hot environments
trichlorethylene	Cold environments
Anesthetic agents:	Noise
epichlorohydrin	Vibration
ethylene dibromide (EDB)	**Stress**
ethylene oxide (EtO)	**Metals:**
formaldehyde	Lead
Rubber manufacturing:	Boron
1,3 butadiene	Manganese
chloroprene	Mercury
ethylene thiourea	Cadmium
Vinyl halides:	Arsenic
vinyl chloride	Antimony
Hormones	**Biological agents:**
Undefined industrial exposures:	Rubella
agricultural work	Cytomegalovirus
laboratory work	Hepatitis B
oil, chemical and atomic work	Other infectious agents
pulp and paper work	Recombinant DNA
textile work	

Source: U.S. Congress, Office of Technology Assessment, Reproductive Health Hazards in the Workplace, U.S. Government Printing Office, Washington, December 1985. p.7.

TABLE 4.2. Known and Suspected Reproductive Hazards by Industry

Artists and Jewelers

boron and boric acid
cadmium and lead
carbon disulfide

Auto Manufacturing and repairs

aromatic hydrocarbons
(benzene,toluene,xylene)
carbon monoxide
chlorinated hydrocarbons
(solvents)
epichlorohydrin
formaldehyde
glycol ethers
heat, extreme
lead and lead compounds
polyvinyl chloride(PVC)

Chemical workers

anesthetic gases
aromatic hydrocarbons
arsenic and compounds
boron and boric acid
chlorinated hydrocarbons
dimethyl sulfate
epichlorohydrin
ethylene dibromide (EDB)
ethylene oxide (EtO)
mercury, formaldehyde
pesticides
selenium

Clothing, textile and leather

arsenic and its compounds
boron and boric acid
carbon disulfide
dimethyl sulfate
dimethylformanide (DMF)
epichlorohydrin
ethylene dibromide (EDB)
ethylene oxide (EtO)
formaldehyde
pharmaceuticals
vinyl chloride (PVC)

Office and other clerical

indoor air pollution
pesticides
solvents, stress
video display terminals

Electrical workers

boron and boric acid
lead or cadmium compounds
PCB's and PBB's

Electronic and semiconductors

aromatic hydrocarbons
(benzene,toluene,xylene)
arsenic and compounds
cadmium and compounds
chlorinated hydrocarbons
glycol ethers, lead

Food workers

chlorinated hydrocarbons
(solvents)
ethylene oxide
heat, extreme
pesticides

General manufacturing

aromatic hydrocarbons
cadmium and compounds
chlorinated hydrocarbons
epichlorohydrin
formaldehyde
glycol ethers
lead and compounds
styrene, vinylchloride

Glass and pottery workers

arsenic and its compounds
boron and boric acid
cadmium and its compounds
lead, manganese
heat, extreme
non-ionizing radiation

Hospital and healthcare

anesthetic gases
carbon disulfide
ethylene oxide (EtO)
formaldehyde
ionizing radiation
mercury

Refinery workers

aromatic hydrocarbons
carbon disulfide
ethyleneimine
lead

Painters

aromatic hydrocarbons
boron and boric acid
lead

Paper workers

chlorinated hydrocarbons
boron and compounds
dioxin, formaldehyde
ethyleneimine
non-ionizing radiation

Pharmaceutical workers

chlorinated hydrocarbons
dimethyl sulfate
epichlorohydrin
estrogens
ethylene dibromide (EDB)
manganese and compounds
mercury and compounds

Plastics workers

cadmium and compounds
dimethylfomanide (DMF)
epichlorohydrin
styrene
polyvinyl chloride(PVC)

Rubber workers

carbon disulfide
chloroprene
lead
manganese

Steel workers

arsenic
cadmium
carbon monoxide
heat, extreme
lead
manganese
solvents

Wood processors

arsenic and compounds
boron and boric acid
ethylene dibromide (EDB)
formaldehyde
mercury and compounds

Source: Adapted from fact sheet, Philaposh, 511 Broad Street, Suite 900, Philadelphia, PA 19123.

THE HEALTH CARE FIELDS

The job of the health care worker is to heal. It is ironic that the process of healing can cause harm, particularly for those working in hospitals and laboratories. They face occupational hazards, some of which cause mutagenic, teratogenic, and carcinogenic changes (see glossary for definitions), illnesses, discomfort, and injury in animals and humans. The jury is still out on many other exposures. The possibilities for harm are so real yet amelioration so meager that in 1986 the New York State Nurses Association received a grant to teach nurses how to protect themselves from on-the-job health and safety hazards.

Health care is also a high stress occupation. Curing and caring provide emotional sustenance to the professional, but illness and death are anxiety-producing events. If you add disagreements among patients, families, doctors, and nurses, the atmosphere can turn into a pressure cooker, particularly in intensive care units and emergency rooms. You become exhausted and can develop headaches, insomnia, ulcers, gastrointestinal problems, high blood pressure, and heart disease – conditions bad for anyone, but even worse for a pregnant woman and her unborn child.

Health care institutions tend to neglect the health of their employees in their attempt to improve the health of their patients, not a very sensible approach. What they save in money, they lose in high staff turnover and burnout.

> I had two miscarriages before my first baby. I was working in the operating room at the time where I was exposed to anesthetic gases. Some of the anesthesiologists had anesthesia machines that were kind of antiquated and there was no way of venting the gases out of the room or they refused to do it because they thought there was no problem. I was really angry at them. They weren't the ones who had the miscarriages and even if it didn't cause a problem, it really wasn't a big deal for them to hook their machines up to the suction equipment.
>
> The hospital did not have any policy and nobody would stand behind me. The nursing director wouldn't and it was easier for me to leave than to get someone to back me. It was easier for me to just transfer out. I didn't have any problem transferring out. It is kind of ridiculous that you have to go to that extreme. I was under a lot of stress in terms of what my environment was causing. I was really upset.
>
> – Clarissa, operating room nurse

The following hazards are fairly common in hospitals. Do you face any of them? They are more serious if you are pregnant and to be avoided as much as possible.

1. ***Injuries.*** These primarily are back injuries and sprained and broken arms and legs. You can try to avoid these by asking for help in lifting heavy patients and loads. Also be extra careful to watch where you are walking, particularly when you are in a hurry, are

harried or stressed. These are the times you are most likely to trip over an electric cord or slip on a wet floor.

2. *Radiation and high energy exposure.* These come from X-ray equipment, particularly portable X-ray machines, radioactive dyes, and implants. Large doses may cause tissue damage and genetic damage. Check with your employer, whether a medical center, health maintenance organization (HMO), private doctor, or dentist—to ensure that a proper maintenance schedule is in effect and that safe handling techniques are included in a required orientation program for all employees. Many workplaces do not meet these standards so you will have to exert pressure through your union, by organizing a group of affected workers, or, in a small practice, asserting your rights yourself.

3. *Clinical hazards.* In the line of duty, certain routine jobs have anything but routine consequences. Sterilizing chemicals, drugs, and anesthetic gases can cause skin and respiratory irritation, liver, kidney, and nervous system damage, cancer, and reproductive dysfunction. You must be extremely careful and wear gloves, lab coats, and masks when warranted.

4. *Infections.* These can vary in severity from the common cold to a staph infection to hepatitis B. The latter is a serious liver disease. In addition, recent research has indicated that a toxic resulting from a hepatitis B infection seems to cause a mutation in a gene which sets off a cancerous growth in the liver. Luckily, there is now a vaccine against hepatitis B that is 90% effective. As of 1988, all health care workers who face routine exposure to blood and body fluids *must* be offered a free hepatitis B vaccine. Workers should see the infection control nurse or health and safety director of their health care setting. The vaccine is not recommended for pregnant women so be sure to be vaccinated *before* you become pregnant.

AIDS is another disease that poses a potential hazard to health care personnel and is of particular concern to pregnant workers, as an HIV positive woman can pass the virus on to her unborn child. It is fortunate that the HIV virus that causes AIDS does not spread by casual contact but only through blood and sexual fluids. So far only a handful of health care workers have been infected with HIV on the job from blood or body fluid exposure. We have been told so often in the past that our worries about potential harm were misplaced, only to learn at a later date that our concerns were justified after all!

This has been true in the past with respect to other occupational health and environmental health hazards. Allowable exposure levels for many toxic chemicals have been shown to have unhealthy effects. In the case of HIV, though, careful investigation of close, non-sexual family contacts over long periods of time have shown that HIV is only transmitted by the exchange of blood or body fluids. In December 1991 OSHA issued a Bloodborne Pathogens Standard designed to protect more than 5.6 million workers. Bloodborne pathogens are microorganisms found in blood that are harmful to humans. They include the hepatitis B virus and the HIV virus which causes AIDS. The standard mandates engineering controls, personal protective equipment, employee training and work practices that will reduce on-the-job risks for all workers exposed to blood. OSHA is issuing a booklet outlining the provisions of the standards which is also available in Spanish.

While the risk of contracting the HIV virus from non-invasive patient contact is minuscule, it is crucial for you to follow the designated procedures set up to protect health care workers from exposure to such a deadly disease. Some of these are:

1. Strict adherence to hospital infection control guidelines regarding sterilization, housekeeping, and disposal of infectious waste.
2. The use of personal protective equipment such as gloves, gowns, masks, and eye coverings wherever contact with blood and body fluids is expected. If body contact occurs by accident, the worker should immediately and thoroughly wash the area.
3. Use of convenient puncture resistant containers for disposal of sharp items and careful handling of sharp items. The ban of needle-breaking equipment.
4. Mouthpieces, resuscitation bags, or other devices to avoid mouth-to-mouth resuscitation should be readily available.

Guides have been written explaining the procedures health care personnel should follow in order to avoid accidental exposure to infected blood or other body fluids. One of the best of these guides, both thorough and easy to read, is published by the Service Employees International Union (SEIU) and costs $2.50 prepaid. The address is: SEIU Health and Safety Department, 1313 L street, N.W., Washington, DC 20005. When you use any guide, check the date of original publication or revision in order to obtain up-to-date information. The AFL-CIO Department of Occupational Safety and Health also has an AIDS in the Workplace Project. For information write to 815 16th St. N.W., Washington, DC 20006.

Some particular health care jobs are more hazardous than others.

The Sterilization Department

Sterilization of hospital facilities and equipment is given high priority. Now high priority has to be given to the sterilizers themselves. Ethylene oxide (EtO) is one of the main sterilizing agents used in the hospital. In laboratory studies EtO has affected both the reproductive capacity of male mice and increased mutations of genetic material in hamster cells. Furthermore, workers in Sweden who were exposed to only low levels of EtO in the process of producing the chemical were found to suffer from leukemia and stomach cancer 10 times as frequently as the national Swedish rates. Even though the use of EtO in hospitals constitutes only about 0.5% of the total production in the United States, health care uses present the major source of human exposure to this chemical.

> I worked in the sterilizing department of the hospital. Ethylene oxide was used for sterilizing instruments, sheets, and blankets in the steam sterilizer. It is a poisonous gas. We wore masks. We also wore buttons to see if gas had escaped. These buttons show how much gas you are exposed to. Gas never escaped while I worked there. The job was physical in the sense that large racks had to be pushed into the room sterilizer and they were extremely heavy.
>
> —Pam, sterilizer of medical instruments

OSHA has set the standard for average level of exposure to EtO as 1 part per million (1 ppm). But in sterilizing material and equipment, workers are exposed to relatively high levels of EtO for brief periods. This usually occurs during the transfer of material from the EtO sterilizer to the aerator unit. Danger of reproductive injury for this short period-high dose contamination typical of the hospital situation is a problem. Many cases of miscarriages have been reported among pregnant women exposed to periodic short-term, high doses of EtO while working with sterilization equipment. It is obvious that a standard for an average level of exposure does not protect you under these conditions. If you are trying to conceive or are pregnant, ask for a transfer to another job if at all possible.

The Cancer Treatment Unit

Until recently, we did not realize the health implications of dispensing cytotoxic anticancer drugs. They are now linked to the development of cancer in the health provider and birth defects in her unborn child. Swedish researchers reported finding increased levels of chromosome abnormalities associated with long-term handling of chemotherapeutic agents. Prior research found that chromosome damage was reversed after measures to limit exposure were installed. A Finnish study found that exposure to cytotoxic drugs may cause the employee to suffer liver damage. Moreover, this type of liver damage is not usually picked up by ordinary liver function tests until the damage is severe. Anecdotal reports by nurses also describe milder effects—facial flushing, lightheadedness, and dizziness.

The National Institute of Health and several other professional groups have issued guidelines for the safe handling of highly toxic chemotherapy agents. In 1986 OSHA also issued guidelines covering the preparation, administration, and waste disposal of cytotoxic anticancer drugs. These recommend established work practices but are not legal requirements. The best advice, especially if you are pregnant, is to be extremely careful and to wear face masks and gloves when handling any cancer chemotherapeutic agents. One word of caution. A report in the *American Journal of Hospital Pharmacy* warned that commonly used material in surgical gloves might be permeable to anticancer drugs. The researchers found that the one commonly used chemotherapeutic agent they tested permeated the glove within five minutes and increased over time. Do not assume that because you are wearing gloves you are completely safe. There are dozens of types of gloves, each protecting you against certain substances. You may have to change to a surgical glove made of different material or double glove (wear two pairs of gloves).

The Operating Room

Several research studies in Europe and the United States have implicated waste anesthetic gases in the operating room in an increase in spontaneous abortions and birth defects among nurses, physicians, and wives of physicians. A nationwide study by the American Society of Anesthesiologists also found an increased risk in women working in the operating room of cancer, kidney, and liver disease. The same linkage was found among dental personnel who used gas anesthetics in oral surgery and rehabilitative dentistry. But other scientists have reviewed the major epidemiological studies and concluded that significant flaws in study design and investigation procedure makes it difficult to justify the conclusion that occupational exposure to anesthetic gases causes increases in miscarriages and birth defects. Thus a controversy about the health effects of exposure to anesthetic gases still exists.

In response to the initial findings, many hospitals improved the scavenger systems to remove most of the anesthesia that had leaked into the air. Not all the newly installed equipment work equally well. You should check and see whether your employer monitors for accidental gas leaks and properly inspects and maintains the equipment and changes the filters.

Often X-ray equipment is used in operating rooms, and pregnant operating room personnel could be exposed to escaping radiation as well as escaping anesthetic vapor.

> Also in the operating rooms that I worked at, they did a lot of procedures that used an X-ray machine in the room throughout the whole procedure. All the circulating nurses and the scrub nurses wore lead aprons for protection. But you know, there is still some radiation. The X-ray techs that came in to do the X rays were always monitored. They wore little badges, but the nurses and doctors were never monitored.
>
> They did not allow people that were pregnant to go in those rooms. However, the two pregnancies that I had were not planned, so how many procedures I did before I even knew I was pregnant, I don't know.
>
> —Clarissa, operating room nurse

The Laboratory

Women of childbearing age make up the majority of clinical and research laboratory workers. Laboratories have many potential work hazards. Some come directly from patients' infections such as hepatitis B and some from the material and techniques used to determine the illnesses. For example, mercury, lead, and carbonate compounds used in laboratory procedures are known to cause birth defects in exposed animals or humans. Dioxane, a dehydrating agent used for slide preparations can cause liver and kidney damage. Radioisotopes and radioactive patient specimens can also be harmful to the unborn child.

How Diseases Are Transferred in the Lab

Factors Involved in Potential Transmission

1. Handling, analyzing and disposing of biological material, e.g., blood, tissue.
2. Inadequate labelling and packaging of biological material.
3. Uncovered or overfilled garbage cans containing carelessly discarded contaminated specimens.
4. Accidental pricking of skin with contaminated instruments.
5. Direct exposure to contaminated blood.

Prevention

1. Use gloves when handling infectious specimens.
2. Do not pipette by mouth. Use disposable pipettes and sterile pipette tips.
3. Wear protective clothing, e.g., lab coats.
4. Use ventilation hoods when mixing cytotoxins or other toxics.
5. Do not eat, drink or smoke in the lab area. Remove lab coat prior to eating, drinking, or smoking in areas designated for those purposes.
6. Clean up all spills quickly and thoroughly with sterilizing solution, e.g., a 10% chlorine bleach solution for blood.
7. Insist that OSHA regulations be posted and enforced. Make sure that your employer provides a complete list of all chemicals used and their toxic effects.

Most of these risks can be prevented by careful work habits combined with stringent enforcement of safety regulations and standards regarding the maintenance of the laboratory itself as well as lab equipment. It is easy to become careless in your routines, and a happy-go-lucky attitude can sometimes ease the stress of a monotonous job. Yet a little extra effort can prevent serious harm to you and your baby.

> In my job I take a risk anyway. We work with blood. It's not always in the bag. You never know when the bag leaks. They were having the AIDS scare while I was pregnant. It was touchy for awhile and hepatitis is a possibility. I'm sure it was a greater risk being pregnant. There was nothing I could do about these risks. I could quit early.
>
> —Maria, laboratory technician

> I had to draw blood from patients up on the floor. I had to work in the lab. I had to work in hematology and in another department called immunology. I didn't take any extra precautions, except when I was drawing blood from patients. The only problem I had with blood was hepatitis as a potential danger. While I was pregnant, I would try to get out of drawing blood from a patient if I knew that patient had a communicable disease. There was usually someone else around who would do it.
>
> I would not go into the room with a patient who had hepatitis. There were a few instances when I was reluctant to go into a room. But I did it, I was just extra careful. I went in with a mask and gloves and gown, and took extra precautions. Ordinarily, I would not have worn gloves or a gown or mask.
>
> It was hard to get close enough to the microscope because my stomach was in the way. The hardest thing physically is to bend over the patient in bed to draw blood because your posture is off when you are pregnant and it's very difficult to straighten up again.
>
> —Jenny, hospital medical technologist

X-ray and Nuclear Medicine Department

Over a million workers are exposed to ionizing radiation (radioactivity) each year and approximately 44% of these are in health care fields. Those of you who work in the X-ray or nuclear medicine department need to be concerned about being accidentally exposed to ionizing radiation. Ionizing radiation is energy that is transmitted in wave or particle form and is capable of causing ionization of atoms or molecules in the irradiated tissue in high doses. It is known to exert strong effects on the developing embryo, fetus, and child as well as affecting normal reproductive functioning in men and women. High doses of ionizing radiation, much higher than doses likely to be found among employees in the medical field, impair testicular function in males. There is also some indirect evidence that it is associated with lowered sex drive and less healthy sperm. So if you are planning to have a baby, the father-to-be needs to be wary of reproductive harm.

Female offspring exposed to ionizing radiation while in the womb can suffer from reproductive disorders years later. These may be abnormalities in their endocrine systems ultimately leading to infertility or abnormal development during puberty. High doses can cause sterility and can initiate menopause. These effects are more common in patients receiving treatment than in professionals treating them. Nevertheless, it is known that if you are exposed to levels of greater than 20 rads while you are pregnant, your baby faces an increased risk of being born with birth defects. Lower exposure from 1 to 10 rads is associated with increased risk of mental retardation, leukemia, and cancer in your offspring. We do not know what effect, if any, extremely low doses of ionizing radiation have on reproduction. X-ray technicians need to be shielded by lead and must remember to take proper precautions at all times.

Make sure that the machinery is maintained properly and that there are no leaks. The portable X-ray machines have the greatest likelihood of leaking and extra special precautions should be taken when working with them. This advice holds true if you are working in a hospital, laboratory, or doctor or dentist's office. Precautions should also be taken if you work with radioactive isotopes or other diagnostic or therapeutic radioactive material. At present, the use of lasers, a form of non-ionizing radiation, as diagnostic and surgical tools has not been shown to be harmful to reproductive functioning.

If you are an X-ray technician, it would be advisable to request a transfer to another job if you are attempting to become pregnant or are very early in your pregnancy. See Chapter 7 for information on the court cases where X-ray technicians won this right.

> The main thing that really scared me was to walk through the radiology department to go to lunch because they have to keep the doors open. I never tried to find out about any of the risks. I would take a different route so that I avoid radiology completely.
>
> —Jenny, hospital medical technologist

The Dental Office

The dental office is a microcosm of the hospital setting, with potential exposure to a combination of chemical, physical, and biological agents. Dentists have been in the news recently because of the first, and only documented, case of HIV infection transmitted from a dentist to a patient. This incident has reopened a discussion of whether all dentists and physicians having direct contact with a patient's body fluids should be tested for AIDS. The medical and public health community are divided over the usefulness of this suggestion. Prior to this, most of the worry has been on the transference of the HIV infection from the patient to the health care professional.

There are over 125,000 dentists practicing in the United States. Only a small percentage are women, but 93% of dentists employ at least one dental assistant or dental hygienist, the majority of whom are women. It is not uncommon for larger practices to hire four or more women in various auxiliary capacities, most of whom are in their childbearing years. These personnel may be exposed to primary herpes and hepatitis, as well as other microorganisms present in the patient's saliva. Waste anesthetic gases, airborne mineral dusts resulting from high speed grinding of dental material, ethylene oxide from sterilizers, mercury from amalgam fillings, ionizing radiation from X rays, high noise levels from drills, and backaches from standing in awkward positions are among the more common hazards.

> When I took X rays I was afraid of the scatter radiation. We tried to work it out that the other girl would take more of the X rays. Sometimes it worked. Sometimes it did not.
>
> —Rosie, dental assistant

Most of the suggestions for improved safety for X-ray and lab workers also applies to dental personnel. A few additional precautions should be taken in the dental office:

1. Obtain a complete patient health history and update it at each visit.
2. Use a rubber dam to limit the spread of aerosolized saliva.
3. Wear a face mask and surgical gloves while working on a patient.
4. Buy handpieces and air-water syringes that can be heat sterilized.
5. Have your patient routinely rinse his/her mouth prior to the start of the dental procedure.
6. Use only X-ray film rated at least speed group d.
7. Have your office inspected periodically by state officials or other qualified experts. If you work in a small office, ask your employer to have regular inspections of all equipment.
8. Leave the room while X-ray machine is operating.

Hospital Laundry and Housekeeping Services

If you are a service worker in the hospital, you probably have not been adequately trained to avoid contamination on your job. Therefore, if you are pregnant, you are probably even more vulnerable to harmful health and reproductive effects than the professional staff who have had some education about the risks and hazards involved in their work. If you are a housekeeper, you should be careful to avoid contact with infectious waste material that is improperly discarded and not labelled as being infectious. If you are a launderer, you should avoid touching contaminated laundry with your bare hands. Too often, infectious materials, instruments, body fluids, and tissues are just thrown into waste baskets or rolled up in laundry. Housekeepers need to inquire whether patients are infectious. If they are, the housekeepers must continue with the strict cleaning procedures designed for infectious areas even after the patients are discharged. Your unions can help you by providing education and training and putting pressure on your employer to establish stricter health safety practices. See further information under laundry and dry cleaning workers in the service sector jobs later in this chapter.

The revised OSHA Federal Hazard Communications Standard (Right-to-Know Law) covers all workers exposed to hazardous substances. This includes those working in health care facilities, whereas the earlier standard only covered workers in the manufacturing sector. It requires all employers to maintain copies of Material Safety Data Sheets (MSDSs) for each hazardous chemical used and to have them readily accessible if an employee requests the information. All employers must provide information and training to all employees assigned to work with hazardous substances. Manufacturers must provide labels for dangerous substances. All employers must develop a written hazard communication program that describes how the hazard communication standard is being met.

Your ability to improve the safety of your workplace will be increased if the regulation is strongly enforced. This is a big "if." Meanwhile urge your professional associations and unions to address your problems and have groups of workers who believe their work conditions are unsafe document the situation.

INDUSTRY

Women work alongside men in virtually every industry. The number of women in each specific occupation is comparatively small. But all together millions of women are involved and they risk potential reproductive injury every day.

> The chemicals there are real strong and I believe that by inhaling them they could give my baby brain damage or something. Or maybe the chemicals there are not good for your skin and they are not good to be inhaling. They had masks but they didn't give them to you. Sometimes they didn't even have the

masks, so you would breathe in this stuff the whole night. But it was dangerous, I believe, to an unborn baby.

—Carla, automobile seat finisher on an assembly line

There was solder smoke. They said that was not really good for you by itself and more being pregnant. We were supposed to get safety glasses and a mask but we never got them. When you cut the wires for soldering they can pop up and hit you in the eye.

—Marcy, solderer and wirer

Altogether more than 50 chemicals commonly used by workers in the United States have been shown to impair reproductive health in animals. We do not yet know how many of these adversely affect humans too. Men as well as women face these risks and both sexes should unite to eliminate these hazards.

The regulatory agencies have been slow to alert workers, set exposure standards, or ban chemicals. For example, it wasn't until the end of 1986 that OSHA finally issued a "preliminary determination" agreeing with the EPA that four chemicals known as glycol ethers may pose reproductive and other health risks to thousands of workers. About 90% of those at risk are painters, printers, furniture finishers, auto body workers, and wood and metal workers. Possible damage to testicles, reduced fertility, fetal abnormalities, and nerve or bone marrow damage can occur at currently allowable exposure levels.

A few industrial jobs are also thought to be associated with increased maternal risk for spontaneous abortion—those involving lead smelter emissions, the manufacture of oral contraceptives, high dose radiation, and "industrial chemical processes." The risk of increased numbers of reproductive difficulties also can occur through paternal exposure to certain substances. Studies have shown that partners of men exposed to vinyl chloride, DBCP (dibromochloropropane), anesthetic gases, and chloroprene have had trouble conceiving, or have had increased rates of miscarriages and birth defects. As we saw in Chapter 1, lead is the best known of the industrial reproductive hazards. Even though a blood lead level low enough to prevent harm to sperm development in males is similar to the level that will prevent damage to fetal development in pregnant women, the effect of lead exposure on men's reproductive health is often not considered.

The Meatpacking Industry

In terms of overall hazards, the meat packing industry (employing about 46,000 women) is the most dangerous place to be employed in the United States. Pregnant women have to work under chronically hazardous conditions. They endure extreme heat or freezing cold. Grease and blood make the floor and tools slippery, and the stench from open animal stomachs and bladders periodically infiltrates the air. The roar of machinery is constant, the speed of the assembly line has been increased, and workers monotonously and repetitively hack away with knives and power saws, injuring themselves and others regularly. Upton Sinclair's description of the brutal Chicago slaughterhouse at the turn of the century is still valid, though now the brutality is encased in modern technology.

Electronics – Silicon Chips

The computer chip industry has been in the news in the past few years because of its Fetal Protection Policies. This industry likes to call its chip fabrication areas "clean rooms" because dust particles are filtered from the circulating air. It was erroneously assumed that these dust-filtering systems introduced to protect the chips would protect the workers as well. For the last decade, the largely non-unionized, young, minority group women who comprise a large segment of the work force in America's (as well as the world's) micro-electronics industry have been complaining about a wide range of work-related health problems due to their exposure to toxic gases and chemicals. These include headaches, dizziness, trouble breathing, allergic reactions, nausea, sore throats, skin irritations, fatigue, and extreme chemical sensitivity. For most of this time, industry pooh-poohed these complaints and attributed them to "contagious hysteria" among the women. It wasn't until some of the male workers developed similar symptoms and non-company doctors investigated the women workers' complaints that it became evident that the "clean" industry was in fact very "dirty."

> When I was working at — there were pregnant women there and they were never told of the risks of the chemicals that they were working with. If I still was there, in fact that's when I did get pregnant and didn't know it at the time, I would be very aware of what the risks are, as far as the chemicals. I would have tried to get out of that department or I would have had to leave work because it is pretty dangerous. But they never said anything.
>
> —Carol, microcircuits assembler

One poorly designed and badly flawed study of a small number of women in a semiconductor plant in Massachusetts found a significant increase in miscarriages among women etching and gas treating the chips. On the basis of this one inadequate piece of research, AT&T banned all pregnant workers from its semiconductor production lines and other companies were pressuring their pregnant workers to leave this work area voluntarily. This was despite evidence that the so called "clean rooms" of the microchip assembly plants were making non-pregnant workers sick as well.

Partly as the result of an intense lobbying and educational campaign by a coalition of women's groups, environmental groups and unions, the Semiconductor Industry Association (SIA) admitted the inadequacy of the study and AT&T rescinded its ban on pregnant workers. In addition, the SIA has announced the formation of a Scientific Advisory Panel to analyze existing knowledge about reproductive and general health in the industry and funded a three and one half million dollar study to research male as well as female reproductive vulnerability and other health effects. (See the fetal protection policy section in Chapter 7 for more information about this kind of discrimination under the guise of protection.)

Seamstresses and Sewing Machine Operators

The garment industry has always depended on female labor and its history has not always been an honorable one. Working conditions are still poor, especially in the non-unionized shops that exploit immigrant and undocumented workers. Working conditions can cause health problems even in the better unionized shops. As was discussed in Chapter 3, sewing and cutting machines often produce high levels of noise and vibrations. Poor ventilation and hot pressing machines can lead to heat stress. In addition to these discomforts, pregnant workers who sew and stitch may be exposed to hazardous chemicals in dyes, synthetic fibers, fabric treatment processes, and especially cleaning solvents.

Workers who sew by hand or by machine may be paying with their health for our thirst for easy-to-care for products. We all own garments that bear the labels "color fast," "permanent press," and "water repellent," but few of us are aware that hazardous chemicals such as biphenyl, acrylic latex, chromium, and methyl-butylketone are involved in these types of processes. Recent research has also linked three categories of dyes—benzidine, o-tolidine, and o-dianisidine—to liver damage and bladder cancer.

The most widely used cleaning solvents used to remove dirt, oil, and grease from fabrics (perc, perchlorethyelene; TCE, trichlorethylene; and trichloroethane) have been found to cause cancer in laboratory animals and if absorbed through the skin or inhaled can cause serious eye, liver, kidney, heart, and nervous system damage. Many of these chemicals may also be capable of causing reproductive harm. Inhalation of synthetic fiber or cotton dust usually leads to respiratory difficulties affecting your health during pregnancy and if severe enough, could hamper your ability to deliver a sufficient supply of oxygen to the fetus.

Women in the garment industry suffer from tenosynovitis, a tendon inflammation resulting from repetitive movements of hands, wrists, or fingers in the same position. Sitting over a sewing machine or standing over a presser for long periods of time can cause muscle strain, and lifting heavy boxes is hard on your back and shoulders. The speed-ups, repetition, heat, and noise combine to produce the stressful working conditions typical of the garment shop.

Stay informed. Use the Federal, state or local Right-to-Know Laws (see appendix at the back of the book) to determine whether you are exposed to any of these toxic chemicals. If you are, request or have your union request that your employer substitute less harmful substances.

THE SERVICE SECTOR

Different occupations within the service sector have many of the same characteristics. The jobs are high-stress and low-paying, have little prestige and no control, are non-unionized, may be part-time or involve evening or weekend hours, require standing on your feet most of the day, expose you to the public, require you to appear cheerful and friendly, and are filled by women. Many of these women are contemplating pregnancy within a few years, are pregnant, or have young children.

In addition, some of the service sector jobs subject you to chemical hazards, indoor air pollution, extremes of temperature, noise, and vibration, heavy lifting and carrying burdens, and physical and social stress. For the pregnant flight attendant, there is the additional risk of being exposed to solar (from the sun) and galactic (from the stars) radiation, particularly on certain routes and during years when there is greater sunspot activity. Pregnant women on the ground are largely shielded by the atmosphere from these two sources of radioactivity. But no shielding has yet been designed that will prevent high-energy radiation from filtering into high flying airplanes. In 1989, the Federal Aviation Administration published data and issued a warning that pregnant women working on long-haul, high-altitude flights over polar routes would expose their fetuses to larger amounts of radiation than federal standards recommended.

By mid-1990, the agency was still evaluating the evidence regarding whether or not further protective action was needed, though it was planning to issue a computer program that could be used by individuals to help figure out for themselves their radiation doses.

Beauticians

Beauticians and cosmetologists face all the problems of other pregnant workers in the service industries cited above. In addition, they are exposed to many potentially toxic substances in organic hair dyes, hair sprays, hair straightening and permanent waving solutions, nail varnishes, plasticizers and resins, and depilatories.

Many beauty parlor workers realize that chemicals that they work with may be harmful, but far fewer are aware of how extremely toxic some of these products may be. In the 1970s a NIOSH study of cosmeticians exposed to hairsprays found that they were more likely to develop chronic lung disease than were a comparable group of non-exposed college students. Organic hair dyes have also been flagged as potential hazards to those manufacturing them, those applying them, and those having their hair dyed. Some scientists have recommended that studies of both birth defects and cancer be conducted among the millions of women using hair dyes as well as of workers who manufacture hair dyes or use them regularly on their jobs.

Continued standing makes varicose veins worse in the pregnant worker. Having your job depend in part on your pleasant manner adds an additional strain to the discomforts of pregnancy, as does working very often in a poorly ventilated and humid salon. Being exposed to chemicals that are known to cause allergic reactions, skin problems, and asthma may also have additional, yet unknown effects.

Because toxic substances can be absorbed through the skin, it is very important to wear gloves when using hair dyes, perming lotions, and other products. The long-term safety of most of these products is not known. In addition, these substances, particularly those applied by spray, linger in the air and can be inhaled. Wearing a protective mask might aid your health, but it is likely to scare your customers away and cost you your job. Therefore, less obtrusive methods must be taken. Now mousses and gels are designed to be directly applied to the hair and can in general be substituted for sprays. Inhalation of the chemicals is lessened, and if you wear gloves while administering the products, the skin route of absorption is eliminated.

While some of the most toxic hair dyes have been removed from the market, the potential hazards of those remaining is unclear, as are the possible human reproductive effects of many of the solvents used in nail beautification products. A new technique of "nail sculpting" involving exposure to toluene, isopropyl alcohol, butyl acetate, ethyl methacrylate, methacrylic acid, and other dusts may have adverse long-term health effects on both the regular customer and sculptor, particularly if she is pregnant. Methacrylate dusts are known mutagens in animals. Some of the animals injected with methacrylates have developed adverse reproductive effects, including fetal death and birth defects. "Sculptors" should work with suction ducts directly over the workbench in order to reduce the fumes and dust levels. High power units are necessary, and it is imperative to replace the filters every one to three months or the units will be ineffective.

Hair International (formerly Master Barbers and Beauticians), 219 Greenwich Road, Charlotte, NC 28211, 704-366-5177, the national professional association, provides some information on standardization of health and sanitation requirements. Pressure from pregnant workers would likely impel the association to raise the issue of reproductive and other toxic exposures on its list of priorities.

Dry Cleaning and Laundry Workers

Chapter 3 discussed the effects of excessive noise and heat and heavy lifting. Dry cleaning and laundry establishments possess the potential for all three problems, which can lead to making pregnancies more difficult.

Laundry and dry cleaning machines are normally operated at high noise levels. If a woman is subjected to this level of noise for a long period of time she can suffer permanent hearing loss. The hot, humid atmosphere of laundries and dry cleaning plants can stress the heart and circulatory system. Dehydration, an effect that is exaggerated in a pregnant woman, is more likely to occur if she is engaged in strenuous physical activities in such an environment. Dehydration interferes with blood circulation to the fetus and may trigger premature labor. A pregnant worker should take particular care not to get overheated, as she has a harder time keeping her body temperature constant because she has to eliminate the extra heat emitted by the fetus. Overheated conditions add to body stress. The pregnant worker, after vacations and weekends off, should be sure to get used to the hot conditions slowly by gradually increasing the number of hours she spends in the hot environment (becoming reacclimated). If the conditions are exceedingly hot, she should avoid this type of work area as much as possible during her pregnancy. See the following list on precautions you can take.

Prevention

1. Install dehumidifiers and air conditioners.
2. Have rest periods in quiet, air-conditioned room where you can sit with your feet elevated.

3. Have a gradual reacclimating schedule after holidays.
4. Cool drinking water should always be available.
5. Use handcreams before, during and after work.
6. Install sound absorbing material on floors, walls and ceilings.
7. Keeping up a good maintenance schedule will reduce noisiness of the machinery.
8. You should be notified if you are sent contaminated laundry to clean and what it is contaminated with.
9. Keep floors dry at all times. This means cleaning up spills promptly. Take no chances of slipping and wear shoes with no-slip soles.
10. Try to alternate lifting with non-lifting jobs. Make sure that you stretch your muscles after lifting to reduce the strain on them.
11. Make sure all particles are vacuumed at least once each day, and more often if the workplace is unusually dusty or contaminated material is being cleaned.
12. Good ventilation and circulation of air is a must.

Lifting and carrying bundles of heavy laundry and working while standing puts a strain on your back and legs. This too is to be avoided during pregnancy, as your back is already strained by the pressure on the lumbar region. Get someone to share the lifting and try to sit for awhile. Slippery floors and escaping steam are also hazards to be avoided.

Waitresses

Waitressing is one of the most stressful types of jobs and waitresses who are pregnant are likely to be subjected to both physical and psychological stress fairly regularly. In fact, a Canadian study found that waitresses, along with nursing assistants, hospital attendants, and certain types of sales personnel, have more spontaneous abortions than would normally be expected. The researchers hypothesized that the high stress level implicit in many of these jobs is partially responsible. It is in the early months when waitresses are adjusting to the changes in pregnancy that they are likely to suffer from the pressure and work conditions the most. They are not likely to work in the last trimester of pregnancy as their bodies become too unwieldy for the tasks required.

Waitresses are rarely unionized and are caught between the demands of their employers and customers. They can work shift schedules or long hours—slow periods characterized by boredom combined with frenetic activity during meal times. During the latter, they run back and forth between kitchen and tables often suffering the tongue lashing of ill-tempered customers and members of the kitchen staff who are themselves stressed. During busy periods, they often do not have time to eat. In some establishments they are given leftover food to take home, some of which may have turned bad from standing out too long.

Exposure to infections, extreme changes in temperature, poor ventilation, and heavy lifting pose additional difficulties for the pregnant waitress. Often during the summer, kitchens are hot and poorly ventilated while the dining area of the restaurant is air conditioned. Waitresses spend most of their shifts going back and forth between workspaces whose temperatures vary drastically. Furthermore many customers smoke while dining, subjecting the waitresses to passive smoke inhalation. They can pick up infectious diseases from their customers. Customers often sneeze or cough into their napkins which the waitress has to pick up along with the dirty dishes.

Constant standing and running around, and the lifting of heavy trays combine to cause back, shoulder, and leg problems. Stretching exercises help, as do support hose and well-fitting, non-skid sole walking shoes. Videotapes and audiotapes of a 50-minute exercise program to music for pregnant women that is approved by the American College of Obstetricians and Gynecologists can be ordered by mail from: Feeling Fine Programs, 3575 Cahuenga Boulevard West, Los Angeles, CA 90068. The video costs $39.95 and the audiotape $14.95. You can either participate in the complete fitness program or concentrate on those exercises that emphasize stretching and flexibility.

The Retail Trade—Sales and Cashiers

Image is of high priority to retailers. Sloth and slovenliness tend to be unforgivable employee traits. Because sales personnel are supposed to appear busy, they may be required to stand at all times they are in public view. An aura of "class" is given to the establishment if saleswomen wear stylish clothing and high heels. Prolonged standing, particularly in high heels, can cause leg pains and varicose veins. Pooling of blood in the legs is a common problem for pregnant women and prolonged standing aggravates the condition. Stress, also part of the saleswoman's lot, is caused by demanding customers, low wages, lack of respect, having to work some evenings and weekends, and the necessity of being continually on view with a smile and polite word for customers.

Continual contact with the public increases exposure to infectious and communicable diseases. It is wise to keep your distance from a customer or customer's child during flu season or when there is an outbreak of German measles, measles, or chicken pox. You can never tell whether some disease they are carrying may pose a threat to your fetus, though the chance of this happening is small. These are the "up front" problems.

Behind the scenes in the stock room, other types of unhealthy conditions prevail. Safety hazards come from boxes stored in aisles and haphazardly placed on shelves so that it is easy to trip over merchandise or have it fall down when you are reaching for something. Back and shoulder strain is caused by lifting, transporting merchandise, and reaching up and down to shelves. Often stores have no windows and you will suffer from indoor air pollution caused by tight building syndrome, as do office workers.

Prevention

1. Take frequent breaks – to sit or to walk around.
2. Stools should be provided as they were in the 1920s.
3. Rotate sitting and standing jobs – by arranging merchandise on the store shelves or in the storeroom.
4. Urge your employer to carpet the floor – reduces leg pains.
5. Have adequate reaching devices – poles, stools, and ladders.
6. Wear comfortable low-heeled, non-skid shoes with support hose.
7. You should be trained in proper lifting techniques.

Women working at checkout counters in supermarkets are increasingly at risk from injury due to the new electronic price scanners. In March 1987, OSHA fined a large supermarket chain stating "employees were exposed to the hazards of cumulative trauma to the back, hands, and arms caused by constant and excessive twisting, stretching, and lifting action while the cashiers lifted, pulled, and pushed merchandise as it was processed through the scanners at the checkout counter." All of these twisting and pulling motions are even harder on the body of the pregnant worker. Several ergonomic improvements were suggested by OSHA that will lower the risk from injury for all employees working with this new technology—soon to become standard in most large stores:

1. Limiting the extent of the arm reaches.
2. Controlling the work height.
3. Providing a "sit-stand" stool for alternate sitting or standing at the counter which allows rest for overworked muscles.
4. Installing right- and left-handed checkout counters.
5. Providing task rotation and work breaks.

AGRICULTURAL WORKERS

The plight of farm workers is among the worst in the country. It is difficult to believe that in 1985 OSHA, after numerous hearings, denied farm workers the right, already enjoyed by all other occupational groups, to have sanitary facilities at the workplace. Some employers failed to provide toilet and hand-washing facilities and drinking water in the field. This was particularly hard on pregnant women who need fluids, access to toilets, and a place to rest out of the sun. This policy was only changed in 1987 as a result of a court ruling. The government has allocated so few inspectors, however, that on many farms the sanitary conditions have not improved and are no better than they were before the court ruling. In a worrisome finding, active cases of tuberculosis were found to be increasing.

About 5 million men, women, and children work in American agriculture and they are regularly exposed to pesticides. In fact, a Department of Health and Human Services study found that one of every three farm workers in southern New Jersey claimed to have been accidentally sprayed by an airplane or tractor rig with unsafe levels of chemical pesticides. This was just in one small section of the country. There is no reason to believe that this is atypical, which means that large numbers of pregnant workers are likely to be exposed to big accidental doses and lower chronic doses of toxic pesticides.

Sprayed workers may experience severe and immediate side effects of pesticide exposure or poisoning such as nausea, occasional vomiting, dizziness, chest pains, eye problems, skin rashes and flu-like symptoms. The health effects of these exposures are further exacerbated by the pressures of working at piece rates, substandard housing without running water, and heat stroke. If you are pregnant and work or live in areas where pesticides are heavily used and you experience any of the above symptoms, inform your local health care provider immediately.

Most cases of pesticide poisonings are not even diagnosed as such. Doctors seldom ask about occupational exposures; the same symptoms can result from a wide variety of other causes. Some mimic normal pregnancy symptoms and are not attributed to the toxic effects of pesticides. Even if such toxic effects are suspected, the worker and farmer often do not know the names of the chemicals to which they are regularly exposed. Frequently, private contractors are hired for pesticide spraying and no one questions the toxicity of the chemicals used.

If you live in an agricultural community, even if you are not a farm worker or pregnant, you should be alert to the possibility of pesticide exposure as aerial spraying drifts with the wind, contaminating backyard vegetable gardens and residents of the surrounding area. If you are pregnant, it may be wise to find out the dates for massive aerial spraying and leave the area for the day or stay indoors with windows shut.

> I didn't really think about it at the time, but I lived where there was a lot of crop dusting. Every winter I'd get a sore throat as a reaction to this cotton defoliant they were spraying around the middle of December. That had nothing to do with the job. It had more to do with where I lived. When I remembered, I would always ask my doctor and he was reassuring even when I was pregnant and due to deliver in January.
>
> —Kitty, teacher in a private school

Further Information

For information about pesticide exposure, call the National Pesticides Telecommunication Network (NPTN) toll free national number: 1-800-858-7378. This is staffed by pesticide specialists at Texas Tech University's Health Sciences Center, School of Medicine and is open to the public 24 hours a day, seven days a week, dispensing information on pesticide products, basic safety practices, health and environmental effects, and cleanup and disposal procedures.

Another and broader source of information about chemical toxicity is *The International Agency for Research on Cancer (IARC) Monographs on the Evaluation of the Carcinogenic Risk of Chemicals to Humans*, IARC Monographs, Vol. 1-29, Supplement 4, International Agency for Research on Cancer, World Health Organization, Geneva, October 1982. The IARC is very concerned with protection of the individual and believes that governments should regard well-established animal carcinogens as also being human carcinogens for the purpose of evaluating criteria to be used for cancer-risk assessment. Because carcinogens and mutagens are highly related, you may be hedging your bet on mutagenic exposure as well, but not on exposure to substances that may cause developmental defects (teratogens).

Many of you may want to do your own general health and reproductive health evaluation of your workplace exposure to chemicals. You will find the following list of questions helpful. Get as much information as you can. You probably cannot answer all the questions and some may not be relevant, but by helping your doctor or midwife in identifying and evaluating possible harm due to workplace hazards, your health care professional will be in a better position to ensure you a healthy pregnancy.

OCCUPATIONAL INFORMATION TO OBTAIN FOR YOUR OBSTETRICIAN, MIDWIFE, OR FAMILY PRACTITIONER

Provide specific information about your current job and past occupations. Include summer, temporary, and part-time work.

A. Job titles

1. Describe exactly what tasks you performed.
2. Describe a typical work shift in detail.
3. Describe the type of workplace and work station.
4. Describe any unusual or overtime tasks.

B. Give starting and finishing dates for each job.

C. Focus on new or changed processes at work.

D. List chemical, physical, biological and psychological stresses at work.

1. Use your rights of access to the employer's medical and exposure records (these rights are granted under the OSHA Medical Access Standard 1910.20).
2. Find out the chemical names of the substances you are exposed to at work. Ask for the MSDS (material safety data sheet) for those chemicals.

E. Try to estimate the extent of your exposure, e.g., your clothes are covered with a fine film of dust an hour after you start work.

F. Provide detailed information about eating, drinking, and smoking in the workplace, i.e., what you and others do, for how long you do them, where you do them, and what work processes are going on concurrently.

G. If you wash or shower at work, describe the facilities and what you do with your clothing.

H. If you wear any protective clothing or hearing protectors, describe the fit, how often you wear them, and how comfortable they are.

I. Attempt to find out from your employer or union whether protective engineering systems and devices such as exhaust and ventilation systems are installed, whether they are functioning, and whether they are adequate.

J. Monitor your symptoms and compare them with co-workers.

1. How soon after you get to work do your symptoms start?
2. How soon after you get home do your symptoms stop?
3. Do your symptoms feel worse when a special process is being performed?
4. Is there a pattern of symptoms among your co-workers?
5. Are there other factors not connected to work that might solely or in combination with workplace exposures be causing your symptoms?

   a. Someone else in the household who may be bringing home a hazardous substance on work clothes.
   b. Some substance in the home such as art material, cleaning fluid, pesticide spray.
   c. Location of home near factory, incinerator, garbage dump, or contaminated source of water.
   d. Use of new cleaning products, soaps, cosmetics, clothing.
   e. A hobby that requires the use of hazardous material.
   f. Smoking cigarettes, cigars or pipes or heavy use of alcohol in the past or present.
   g. Previous change of residence because of a health problem.

Source: excerpted from Levy, Barry S., "Recognizing and Preventing Hazards in the Workplace," *Consultant*, November, 1983: 63-78, and "The Occupational History," Occupational Health Committee, *Annals of Internal Medicine*, November, 1983, Vol. 99, No. 8, 643-644.

SUGGESTED READINGS

Barlow, S. M. and Sullivan, F. M. 1982, *Reproductive Hazards of Industrial Chemicals: An Evaluation of Animal and Human Data*, Academic Press, London.

Chavkin, W., ed., 1984, *Double Exposure: Women's Health Hazards on the Job and at Home*, Monthly Review Press, New York.

Hricko, A. and Brunt, M. 1976, *Working for Your Life: A Woman's Guide to Job Health Hazards*, Labor Occupational Health Program, University of California, Berkeley.

Irwin, M., 1986, *Risks to Health and Safety on the Job*, Public Affairs Pamphlet No. 644, Public Affairs Committee, New York.

NIOSH Morbidity and Mortality Weekly Report Reprint, 1985, *Prevention of*

Leading Work-Related Diseases and Injuries, 34, September, U.S. Department of Health and Human Services, U.S. Government Printing Office, Washington.

Rogan, W. J., 1986, "Breastfeeding in the Workplace," *Occupational Medicine*, July-September, 411-413.

Stein, Z. and Hatch, M. eds., 1986, *Reproductive Problems in the Workplace, State of the Art Reviews*, Occupational Medicine, Vol. 1, July-September, Hanley and Belfus, Philadelphia.

U.S. Congress, Office of Technology Assessment, 1985, *Reproductive Health Hazards in the Workplace*, U.S. Government Printing Office, Washington.

Photo © Earl Dotter

Chapter 5

Women in the Office, in Schools, and at Home

Office workers, schoolteachers, and homemakers suffer from more workplace hazards than most people think. Stress and indoor pollution are the two main culprits, hard to pinpoint and hard to ignore. Indoor air pollution has reached crisis proportions and millions of women of childbearing age spend large portions of their lives in these contaminated environments. Even sitting in one place for a long time can cause harm.

> If standing all day at work in an overheated factory causes tiredness of the muscles and also varicose veins, prolonged sitting may be just as harmful, for the lumbar region of the spinal column becomes bent, the movements of the abdominal viscera are interfered with, the lower ribs are compressed, and since deep inspiration is hardly possible the lungs are badly ventilated and the aeration of the blood is imperfect.
>
> Thomas Oliver, *Diseases of Occupation*, Methuen, London, 1908, p. 11.

Even though evidence was available that prolonged sitting could cause harmful health effects 100 years ago, until recently women in white collar positions who complained of headaches, dizziness, coughs, colds, nausea, generalized weakness, joint and muscle pain, and various allergic reactions were thought to be "hysterical," a term applied to women when males in charge cannot isolate an "objective" cause. Now it is thought that a combination of sitting in one position, stress, and sensitivity to chemical pollutants in the indoor environment may account for these symptoms. Office work is a "high-stress" occupation and pollutants in airtight office buildings appear to be a major cause of ecological illness (chemically/environmentally induced disease, also referred to as multiple, chemical sensitivity).

Formaldehyde, for example, is considered to be a prime culprit of allergic sensitivity, as well as a carcinogen and possibly a reproductive hazard. It poses a great potential threat to our health as it is omnipresent – found in pressed wood products making up a large percentage of furniture and building materials. Even harder to avoid is its presence in some office supplies, tobacco smoke, perfume, toilet paper, permanent press clothing, carpeting, and drapes. Despite accumulating evidence that formaldehyde posed a danger to workers, it took a seven-year battle waged by unions and the American Public Health Association to persuade OSHA to reduce permissible workplace levels of formaldehyde to 1 part per million. Unless a building is properly ventilated, pollutants build up to levels that are unhealthy and possibly dangerous to you and your unborn children. Because you hyperventilate during pregnancy in order to obtain the oxygen, your total exposure to pollutants increases.

Most ventilation systems in closed buildings primarily re-circulate 80 to 85% of the same indoor air already polluted with tobacco smoke, microorganisms, chemicals from copying machines, cleaning fluids, and synthetic clothing, furniture, and building products. This is mixed with 15 to 20% fresh air (which also may contain pollutants). Unless the system is correctly designed and in very good working order, it will not be able to eliminate even moderate amounts of pollutants. Because you cannot open the windows, you are captives of the heating and air conditioning engineering control systems which do not manage to keep workers comfortable much of the time. Offices, factories, hospitals, schools, and homes can be plagued by tight-building health hazards.

> From what I understand we work in a condemned building. The engineer said that there was a crack down the main seam of the building. But there have been no cave-ins. But the fire department, I believe, won't pass it for inspection. But we're still working in it. There was bad air circulation. When you are pregnant the baby cuts your air anyway especially the bigger

you get. Then if the building you're working in is not really properly ventilated, it puts a strain on you more than you realize.

—Daisy, principal clerk

The air was very bad. In the winter and the summer or all year round it was very bad. It seemed like you were suffocating. Some days would be really hot and it seemed like I was going to pass out. This was in the summer time. We don't have any windows at all.

—Judith, clerk-typist

HAZARDS IN THE OFFICE

VDTs: Stress on the Workfront

In addition to indoor air pollution, the automation of the office has had considerable physical and psychological effects on office workers and has received an enormous amount of media attention. Over 10 million people, 5 million of them women of childbearing age, make their living doing office work. For those whose jobs are varied so they only use VDTs part-time, VDTs have been a blessing, simplifying the tedious aspects of their jobs and increasing the time spent on the more creative or decision-making aspects of their work. But most women do not hold these kinds of jobs. Instead, their pleasant working conditions have been turned into white collar assembly line positions.

Work is now alienating and demeaning. The VDT is master. It can automatically pace your work by bringing new tasks onto the screen and it can supervise and monitor tasks by registering mistakes. The combination of high demands with low control is the hallmark of very stressful jobs. Nowhere is this more true than with clerical VDT workers. In fact a NIOSH survey of clerical video display operators found high levels of anxiety, depression, confusion, fatigue, and health problems. They suffered from even greater work stress than air traffic control operators. At least air traffic controllers deal with potential life and death situations which are of public concern.

While most office workers are still not organized, Nine to Five, the Service Employees International Union (SEIU) and the Communication Workers of America (CWA) have made tremendous strides in organizing and gaining better working conditions. The advent of the VDT made the climate ripe for white-collar unionization. The following list of complaints make it easy to see why.

1. Possible miscarriages, birth defects
2. Soreness, itching, and general discomfort of the eyes
3. Pains in neck, back, shoulders, arms, and fingers
4. Headaches
5. Blurred vision and cataracts
6. Dizziness and nausea
7. Rashes
8. Stress-related illness and discomfort
   a. Irregular menstrual periods
   b. High blood pressure
   c. Inability to sleep
   d. Tension and fatigue
   e. Ulcers
   f. Inability to relax without drugs, cigarettes, or alcohol
   g. Poor appetite or incessant eating
   h. Depression
   i. Frequent sickness
   j. Chronic anger

Glare reducing devices, moveable keyboards, amber and multicolor screens, special glasses, and more frequent breaks are now believed to help. Stretching exercises performed at the workstation also alleviate stress and strain. These promote relaxation and flexibility and can usually be easily performed in the early months of pregnancy. If you are starting these exercises for the first time when you are pregnant, check with your doctor before you begin just to be sure that you do not have a condition that contraindicates their use. Use your judgment and eliminate those specific exercises that might cause you to strain yourself.

Possible Reproductive Hazards

Most of us have read articles about clusters of VDT operators who had miscarriages, premature births, or infants with birth defects. So have women all over the world and they are vocally expressing their anxiety. Yet so far scientists have not been able to determine any direct causal connection between VDT use and reproductive harm. Some scientists view these clusters as clues to a serious problem, while others consider them to be due to chance. The latter group reason that miscarriages are such common events that these clusters could occur by chance alone in the large numbers of women of childbearing age working on VDTs all over the world.

One recent study conducted at the Kaiser-Permanente Medical Center in Oakland, California found that women who used video display terminals for more than 20 hours each week in the first three months of pregnancy had nearly twice as many miscarriages as women doing other types of office work. But this difference was only found in some job classifications and not in others. The researchers also found that heavy VDT users were more likely to give birth to babies with birth defects, but the increase was not statistically significant. However, the members of the Kaiser-Permanente research team emphasize that they cannot be sure that the rise in miscarriages and birth defects was due to the computer itself, or to some physical or stressful aspect of the workplace, or even to some socio-economic differences between those who use the computer most of the time and those who use it less frequently.

Workers have not been satisfied with the negative or ambiguous findings of previous research, and primarily as a response to worker pressure, scientists are investigating low-level radiation and magnetic field effects of VDTs. Some experts suggest that low-level electromagnetic radiation may alter or disrupt embryonic or fetal cell development. Experiments with mice and chicks have shown these effects. Other scientists claim that the designs of earlier studies underestimated the link between VDT use and pregnancy complications.

Information about fetal damage from other sources of radiation indicates that the most critical period for possible harm occurs during the first five weeks of pregnancy. Furthermore, the potential risks of VDT work for men has not yet been identified. Given this amount of uncertainty, a policy that allows both men and women trying to conceive to transfer temporarily to other clerical jobs without penalty might be initiated. Until the data on possible reproductive harm is complete, jobs should be designed so that women can avoid intensive close contact with VDTs. This can be partially achieved by not working near the back of another operator's video display terminal and spending only a limited number of hours per day at a terminal.

Some unions have followed this line of reasoning and have negotiated agreements with employers giving pregnant women the right to perform other equivalent work not involving the use of a VDT during pregnancy, or to take pregnancy leave without loss of seniority rights.

> A lot of jobs are going from manual to computerized systems. I found out that they do give off radiation but only to a certain extent if you sit there for a certain period of time. But, like I said, just knowing they gave off radiation was enough for me. I didn't think there was a risk on my job as long as I wasn't doing the work on the CRTs (video display terminals), which I wasn't. When I told them how I felt about it and didn't want to use the CRTs, they said, "fine."
>
> —Daisy, principal clerk

Nine to Five, the National Association for Working Women, has publicized many clusters of birth defects reported among VDT operators. Presenting this type of information before the cause is ascertained has political and economic repercussions. What measures to take in the face of uncertainty comes up over and over again, especially when scientists are not positive that risks high enough to warrant extensive changes do exist, as is the case with VDT work and reproductive harm. On the other hand, if preventive measures are not taken immediately and the risks do turn out to be high, many people's health will suffer during the interim period it takes for scientists to become certain of the connection between work exposure and reproductive harm. Much information about occupational reproductive health hazards never reaches the public at all unless health activists continually lobby for its release.

By giving the possible effects on pregnancy a great deal of publicity, Nine to Five and the SEIU (Service Employees International Union) played a major role in successfully persuading the National Institute for Occupational Health and Safety (NIOSH) to conduct an epidemiological study of the possible reproductive hazards of VDT work. The results of this 6-year study of 730 telephone operators who had been pregnant at least once between 1983 and 1986 was released in March 1991. The proportions of live births, still births, and spontaneous abortions were similar for those operators who worked on computers and those who did not. The two groups worked the same amount of time, had the same work practices and breaks, were similar in age, number of pregnancies, race, education, and years employed. This study design has been criticized because it is a retrospective study that relies on the memories of the workers, does not address the relationship between work stress and miscarriages, and does not measure individual dose measures.

Another epidemiological study of VDT work and spontaneous abortions with better matched control groups is being conducted by researchers at the Division of Occupational and Environmental Medicine at Mount Sinai Medical School in New York City. In addition, future studies planned by the government will focus on the relationship between hours spent at video terminals and premature births, low birth weight, and birth defects. The final answer on the connection between VDTs or the kind of VDT work and reproductive hazards is still not known. It is only after consistent results are found in several well-designed epidemiological studies that women can be relatively sure about the reproductive implications of VDT work.

If you have or know someone who has an office/work-related pregnancy problem, contact:

Campaign for VDT Safety
c/o 9 to 5
1224 Huron Ave.
Cleveland, OH 44115

This is a joint effort by 9 to 5 and the Service Employees International Union (SEIU). Also contact:

Occupational Safety and Health Office
Communications Workers of America (CWA)
1925 K Street, N.W.
Washington, DC 20006

In order for you to make up your own mind about how much of a risk your individual VDT work situation poses, you need to understand the components of a VDT, how they operate, what the known adverse health effects are, and the basis for suspicions of possible additional types of harm. Remember, being in good physical shape during pregnancy, as well as avoiding specific reproductive hazards, is important for your own health and that of your unborn child.

Heat

VDTs produce about 300 to 400 watts of heat output – somewhat more than electric typewriters. This heat is not a hazard by itself, but large numbers of VDTs combined with poor ventilation can raise room temperature to a level of discomfort. Because you have to cool your fetus as well as your own body, you will be made even more uncomfortable when you are pregnant. Heat output should be measured near the screen as well as other places in the environment as the VDT operator's mucous membranes in the nose and eyes may dry out if the surface temperature of the VDT is too high. Fans can be used to ventilate VDTs, but these fans can cause unpleasant drafts which can result in muscular pains. You can check the temperature on the surface of the VDT after two hours of continuous operation. If it exceeds 90 degrees fahrenheit, bring it to the attention of your supervisor.

Noise

The noise from a VDT can come from ultrasound or high-pitched noises from electrical components, the fan, disk drives, printers, and other adjunct equipment. This noise is seldom hazardous to hearing but it can be annoying and stressful.

Radiation

The cathode ray tube can produce a certain amount of *ionizing* radiation (X-ray) and various electronic components can emit *non-ionizing* radiation (radio frequency). The ionizing radiation produced is small and is almost entirely absorbed by the thick glass screen. The non-ionizing radiation can be reduced greatly by the use of a simple metal shield around the body of the monitor. Women should press for the installation of these shields on the current generation of VDTs and for the development and use of a non-cathode ray tube VDT. Non-ionizing radiation has been inadequately studied in humans. The following harmful reproductive effects have been found in animal studies:

1. High dose microwave and radio frequency radiation (in animal studies):
   a. Chromosome damage
   b. Reduction in sperm count
   c. Cataracts
   d. Blood and nervous system disorders
2. Low dose microwave and radio frequency radiation
   a. Lowered fertility, death of embryos, increase in fetal malformations, and still births in animals
3. High dose ELF (extremely low frequency radiation)
   a. Lower birth rates
   b. Birth defects
   c. Female reproductive disturbances
   d. Increased offspring death in animal studies
4. Low dose ELF
   a. Abnormal embryonic development

So far, tests in the U.S., Canada, and the U.K. show that the radiation from VDTs is extremely low—usually far below the permissible occupational health level. While the results of these tests suggest that there are no significant radiation hazards from VDTs, there is still reason to be cautious. Older VDTs, VDTs that are badly maintained, and VDTs not made according to strict safety standards enforced in most industrialized countries may not be as safe as those tested. The occupational exposure levels also may be too high and the effects of long-term exposure to low levels of non-ionizing radiation are still mainly unknown.

More importantly, as yet there are no federal standards to protect workers from non-ionizing radiation in VDTs – ranging from radio frequency, very low frequency (VLF) and extremely low frequency (ELF) radiation. Most of the non-ionizing radiation emitted by VDTs occurs in the VLF range. Almost no studies have been done at these frequencies. But biological effects have been found at both higher and lower frequencies. The list below summarizes recommendations for manufacturing, maintaining, and operating VDTs.

1. Manufacturing or refitting with protective metal shields to reduce non-ionizing radiation emission.
2. Adjustable screens and keyboards with glare reduction devices.
3. Mandatory, periodic testing of equipment.
4. Scheduled rest breaks – 15 minutes off for every hour on the machine. Maximum 4 to 5 hours of VDT work a day.
5. Elimination of machine-pacing or computer-monitoring of workers' output.
6. Mandatory visual testing of operators.
7. Workers should keep track of symptoms they and co-workers have and record all incidents of machine breakdowns.
8. Worker participation regarding use of VDTs, including job security, work station design, maintenance, job rotation, and any performance-monitoring programs.
9. Ergonomic requirements:
   a. Adjustable table that allows leg movement and at least 7 inches of knee clearance.
   b. A foot rest for short operators.
   c. A simple, movable document holder.
   d. A maximum viewing distance, eyes to screen, of 27 1/2 inches.

Again the U.S. is behind other countries. The European Community issued a Video Display Unit Directive covering most of the points which is to be implemented in the 12 member nations by the end of 1992. However, OSHA is considering whether or not to pursue rule-making in the area of ergonomics.

Other Office Equipment

Next to VDTs, photocopying machines are the main staple of modern offices. When the photocopying machine breaks down, so does the office system. Some of you spend much of your day photocopying memos and reports. Others only use it occasionally. Even bosses run out and copy a few pages themselves. What can be harmful about such a work-saving machine? Surely it is better than using reams and reams of carbon paper. True, but there are hazards with copiers, particularly if they are old, not maintained well, and are run continuously in an inadequately ventilated space. Photocopiers usually operate by an electrostatic process and produce ozone, a sweet-smelling gas formed from the oxygen in the air when it has been energized by the high-voltage source in the machine. Ozone can impair your lung function, which in turn can reduce your resistance to disease. It is also suspected of being a mutagen causing genetic damage. In light of these suspicions, it makes sense not to sit in the path of the photocopier exhaust. Most photocopiers manufactured today filter out the ozone produced.

Another hazard of photocopying machines is the chemical composition of the toners used in the machines. In the past, toner, composed primarily of carbon black, may have been contaminated with polycyclic aromatic hydrocarbons (PAHs), also found in diesel exhaust. Studies on firefighters who have been exposed to these chemicals while fighting fires indicate a possible association with lung cancer. Carbon black used in toners today, however, is generally free of these contaminants and the newer machines have toner recovery systems so that far less toner becomes airborne. Exposure to users is usually minimal, but could be occasionally serious for maintenance personnel. Sometimes one office worker is in charge of changing the toner. If you happen to be the one, try not to touch the toner with your bare hands or breathe it directly. While the chemical contamination problem has largely been eliminated, photocopiers, in general, dry and ionize the air, which causes discomfort to some workers.

Carbonless carbon paper, another boon to the office worker, also has its darker side as some carbonless carbon paper has formaldehyde as one of its ingredients. A few workers become sensitized to formaldehyde, and others find the paper irritating to their nasal membranes, bronchial tubes, and skin.

The principal solution to the ozone problem as well as other indoor air pollution problems is better ventilation. Most manufacturers have minimum space and ventilation requirements for the machines they sell. Check the manual for the equipment you have or call the rental company for this information. If your location falls below the standard required, either move your machine to a safer location or rent a different type of photocopier that fits your environment.

HAZARDS IN THE SCHOOLS

Teaching, particularly at the pre-school and primary school level, has traditionally been a woman's job and still is to a great extent. Women did not have to worry as much about reproductive health hazards in the past. They were forced to quit their jobs either when they married or when they became pregnant. Now that these barriers have been withdrawn, thousands of pregnant women teach through most or all of their pregnancies at every level of the educational system. Yet school administrators have done little to protect their health and the health of their unborn children.

While schools are safer places to work than many manufacturing plants, teaching is not a hazard-free occupation. Pregnant as well as non-pregnant teachers face risks from asbestos crumbling from the ceiling, discipline problems involving physical violence, high noise levels, childhood disease epidemics, insufficient opportunity to use restrooms, old copying machines used in airless rooms, unsafe laboratory procedures, and toxic art supplies.

> There were bugs in the bathroom, a set of slippery staircases, and a room that they were removing asbestos from. I had to stay out of that room. I was concerned about keeping warm and dry on recess duty. You could use the bathroom

whenever necessary as long as there was a teacher available to watch the kids. Sometimes there wasn't though.

—Dinah, elementary school teacher

I would pick up clients and take them to bowling tournaments. Sometimes we had clients who were disruptive, abusive, or runaways. These were hazardous conditions because you could have actually been hurt. In fact, in one incident, I had a client who refused to turn off his T.V. When I asked him to turn it off, he went off. He started kicking me in the legs and pushing me and at that time I was 8 or 8 1/2 months pregnant.

If a staff member was hurt, they would have to wait for someone to come relieve them before they could go get medical attention. There should have been someone working with me because I didn't know when I was going into labor. If you couldn't get to a phone you had to lie there until another tour group came along. One day I fainted. I lay there until I recovered and continued the tour.

—Penny, recreational counselor at a residential center for individuals who were retarded or had special problems

Art Teachers

If you are a pregnant art teacher, you are likely exposed to more toxics than your non-pregnant colleagues or students. Because your respiratory rate has increased, you may inhale greater quantities of toxic substances from the air. Substitution of products, careful handling of material, and adequate ventilation will reduce your exposure level substantially. For example, common solvents such as toluene and carbon tetrachloride used by sculptors, printmakers, and painters in mixing paints, varnishing, and cleaning are very toxic. Carbon tetrachloride is essentially banned today as it is a potent liver carcinogen and should be disposed of immediately in accordance with the safety laws. Right-to-Know laws require inventories and removal of dangerous substances. Schools, however, are sometimes negligent in taking inventories and new safety regulations often only slowly trickle down to the teachers. A good resource on substitutes is *Artists Beware* by Mike McCann.

Some paints, solders, and dyes are toxic to the kidneys, liver, and lungs. Some can cause cancer and damage male and female reproductive systems. Beware of those whose ingredients include heavy metals—cadmium, chromium, lead, mercury, and manganese—suspected carcinogens and mutagens. In addition, clays, glazes, and sculpting stones can contain silicates. Mixing, grinding, and carving processes create silica dusts which can cause silicosis, a lung disease. Firing processes also can cause the formation of toxic gases. Aside from the chemical ingredients, poor maintenance and lack of safety features on equipment and insufficient circulation of fresh air can lead to increased hazardous exposure. The following list offers suggestions on how to protect yourself.

What You Can Do

1. Read all labels carefully. Unfortunately, too many art supplies provide insufficient information. Don't take any chances, particularly if you are pregnant. If in doubt contact:

   Art Hazards Project
   Center for Occupational Hazards
   5 Beekman Street
   New York, NY 10038
   (212) 227-6220

   This project answers telephone and written requests for information and has published a booklet about reproductive hazards in the arts and crafts which includes a brief discussion of how toxic substances can affect reproduction, steps to take to avoid the hazards and a list of known effects. It is available from the Center for Occupational Hazards in the Arts at the above address.
2. Make sure all containers are tightly covered and stored in safe places.
3. Wear protective clothing and goggles when required.
4. Vacuum dust and particles. Sweeping only spreads them in the air and onto your clothes.
5. Do not eat, smoke, or drink in your work area.

6. Make sure that the art classroom is properly ventilated.
7. Do not mix liquids while disposing of them. Throw away rags and towels used for cleaning. Do not leave them lying around or in the wastepaper basket.
8. Follow these rules at home as well as in the art room. The same toxic ingredients in inadequately ventilated rooms, improperly handled, will cause the same harm regardless of location.

Science Labs

Science labs are another hazardous area for the pregnant teacher. In school laboratories the hazards of working with students are substituted for the hazards of working with patients. You must take extra care especially because students tend to be nonchalant about safety precautions. Potentially explosive, carcinogenic, toxic, and irritant chemicals are being used. Usually the students and sometimes even less qualified teachers are unaware of proper precautionary procedures. In 1984, the Council of State Science supervisors prepared a guide addressing these safety issues. It lists potentially hazardous chemical substitutions, as well as information about the safe storage of chemicals. In addition, under the Right-to-Know laws, if you are a science teacher, you should receive information and training about the safe use of hazardous chemicals. In 1990, a new OSHA regulation went into effect called Occupational Exposures to Hazardous Chemicals in Laboratories which applies to all school and research labs. It requires a chemical inventory, worker training, appointing chemical hygiene officers, and proper work practices. The CUNY Center for Occupational and Environmental Health (listed in the resource section at the back of the book) has developed a guide for complying with the OSHA Laboratory Standard which you can obtain for a minimal photocopying fee.

Along with being adequately informed about the toxicity of the agents and chemicals used in the labs, you and the students are urged to wear protective clothing and closed shoes. Shorts, miniskirts, and sandals have no place in the science classroom on experiment days regardless of students' possible cries of outrage.

Duplicating Machines

Duplicating and copying machines in the teachers' lounge pose the same hazards that they do in the office (see earlier section of this chapter on office machines). Most often teachers use mimeographing and duplicating machines requiring stencils and inking of drums because they are much cheaper than the more modern copiers. What aggravates the problem even more is that the machines tend to be older and may not be regularly maintained. Teachers often poke and pull in order to get the old mimeographing machine operating or the old duplicating machine with the smelly wet copies to print clearly.

You are more likely to get chemicals all over your hands if you are rushed. Still more dangerous is that these machines are frequently located in your lounge where you and your colleagues rest and eat. Thus you are apt to munch while you copy, ingesting the chemicals along with your lunch and adding pollutants to others' lunches as well.

> There was a teachers' lounge which also had the xerox and the ditto machine. We ate in there and any lounging was done in that room.
>
> –Kathy, second grade teacher in a private school

Infections

Any job that brings you in close working contact with the public brings you in contact with many different kinds of germs. Pregnant teachers who work with younger children are particularly at risk because this is the age group that becomes infected with all the childhood diseases. Those who evaded the contagious diseases when they were children may catch them from their pupils. Adults usually have more serious cases. For example, German measles (rubella) is a known human teratogen, and all female teachers contemplating pregnancy in the future should be vaccinated against the disease if they do not already have immunity due to previous exposure. Male teachers who plan to start a family or have additional children are also vulnerable. Severe cases of mumps can cause sterility.

Pets can breed disease, and pregnant teachers should be wary of sick hamsters and gerbils. Children can catch toxoplasmosis disease from their animals and bring it to school. Toxoplasmosis can be transmitted through the placenta possibly causing damage to the nervous system and eyes of the fetus. In addition, with more mothers working, more children are being sent to school sneezing and coughing, whereas in former days they would have been kept at home. Most diseases are infectious just before the child comes down with it and this is the time the child is likely to be in school. Mothers become skillful at knowing when their children are faking, but they are not infallible in separating complaints of not feeling well due to illness from those due to not wanting to go to school.

Because more and more mothers are working, and more and more children are attending day care centers, more and more day care personnel are needed. This new cadre of women are even more at risk for stress overload and infections than nursery school or primary school teachers. These very young children may not be toilet trained nor have they yet developed good cleanliness habits. Diarrheal infections, including the parasite *Giardia lamblia*, hepatitis-A as well as upper respiratory infections, rubella, and cytomegalovirus have been known to spread through day care centers.

Overexertion

Overexertion from lifting is another chronic hazard for the pregnant day care or nursery school teacher. Lifting heavy children puts just as much strain on your body as the lifting of heavy objects and can also cause a miscarriage. The same lack of consideration concerning the needs of pregnant workers is found in schools as is found in offices, factories, and health care facilities.

> I lifted 35-pound children and retrieved kids that climbed too high on the playground equipment. There was a lot of bending and from September to November I had contractions every time I would bend over. The kids would just wait for them to finish and then they could keep going.
>
> I had no break for a 5-hour period. There was no snack bar with nutritious food. We just had whatever we fed the kids which was not necessarily nutritious. There was a kitchen that

we could use where we could store sodas or juice, but we couldn't have it until lunch or work was over.

If we had a circle to read the children a story, we had to be on the floor with them. At the end of my pregnancy, I asked if I could please use a chair because to get down on the floor and to get off the floor, stand up and down, jumping around when you are 8 months pregnant was really hard. They told me begrudgingly that I could do that. If I had time to actually sit and elevate my feet that would have really helped.

—Amy, day care teacher

NO LONGER A HAVEN: HAZARDS IN THE HOME

Lead Paint

A woman's haven may be her home, but increasingly it is a polluted one. While we realize that we may face hazards in our workplace, we tend to think of our homes as sanctuaries. It comes as a shock to find out how wrong we are. The hazards come from such diverse sources as the vinyl that is used in the manufacture of furniture and flooring today, to the lead paint found on the walls of old houses. While Congress banned the use of lead-based paint more than ten years ago, a Government study released in 1990 reported that about 57 million homes—two-thirds of those constructed before 1980—still contained this type of paint. The older the home, the more likely there is lead paint. The lead paint is not only found in slum tenements, but in expensive housing as well. The director of the Federal Centers for Disease Control believes that lead poisoning is the number one environmental hazard facing American children. The biggest threat seems to be from lead dust, rather than from chipping or peeling paint. Severe retardation, stunted growth, lowered intelligence, hyperactivity, subtle problems in early childhood development, and, in the rare instance, even death have been linked to lead eaten or inhaled by young children. Massachusetts is one of the few states that has strict laws requiring paint removal. We know that lead is detrimental to the fetus, so where does that leave the pregnant woman living in a house or apartment whose walls still are covered with this toxic substance?

If you are pregnant, there are a few things you can do to help protect yourself as well as your young children. First, remember that if lead paint is bound to the surface, it is hard to get into your blood. If there are several layers of non-lead paint on top of the older toxic paint and there is no peeling, there is probably no problem. Experts recommend avoiding partial, do-it-yourself paint removal projects in older homes. These weekend projects tend to drag on and end up exposing layers of lead paint that had previously been safely covered and releasing lead dust. The most important thing you can do is to keep your walls, woodwork, sills, and floors clean, as routine washing will greatly reduce the risk of exposure. The most hazardous areas of exposure are stair banisters, old moldings, window sills, and woodwork that your child can reach. If these are covered with cracking or peeling paint, you might want to consider replacing them if you are worried about the possibility of releasing lead.

If you are worried that you or your child may be suffering from lead poisoning, ask your doctor to test your lead blood level. The lead poisoning prevention branch of the Centers for Disease Control is expected to recommend that physicians routinely screen children under seven years of age for lead poisoning. The agency may also lower the exposure level that is considered hazardous from 25 micrograms per deciliter of blood to 15 or 10 micrograms. This will greatly increase the number of children thought to be at risk.

Electromagnetic Fields

Even electromagnetic fields from power lines near or in your home have been identified as possible health hazards. In 1990 a report by EPA scientists concluded that these fields were a possible, but not probable cause of cancer in humans. Furthermore, researchers who exposed laboratory animals to powerful electromagnetic fields found that their offspring were smaller and had a higher incidence of birth defects than those not exposed. These birth defects seemed to be passed on to ensuing generations.

At present, there is not enough evidence to warrant doing anything about electromagnetic fields in the home. They may very well turn out to be too weak to wreak any havoc. But be alert to future scientific evidence about this potential hazard. If you are particularly worried, there are two possible ways of eliminating, or reducing, the problem. Switch from copper plumbing pipes to plastic ones. The latter cannot carry an electrical current. Also have the electricity coming into your house leave it on an adjacent line. Electromagnetic fields from two currents running in opposite directions will cancel each other out.

Home Work

Another potential problem is what is referred to as the new cottage industry or "home work." As women find it more and more difficult to find adequate and affordable child care, they have begun to work at home. They may do word processing or work in the garment industry. If they are artists or craftswomen, professionals who are on a regular salary, who work home one or two days a week, or are well-paid consultants, this usually works out fine. But if they are paid on a piecework basis or their keypunching is monitored, they may be repeating the story of the poor, exploited immigrant women at the turn of the century. Without union backing to achieve a living wage, and without medical insurance or fringe benefits, today's poor mothers are being exploited once again. They suffer from stress, backaches, headaches, eyesight problems, cumulative strain injury, or worse. In other words, they suffer from all the problems other workers suffer from, but with far less, or no, protection. If they are craftswomen, they probably do not have the proper air filtering system and toxic chemicals may be circulating in the air. This will worsen the already existing indoor air pollution.

Air Pollution

The quality of the air in our homes has become a major problem, and it is no longer clear that it is always healthier to stay inside when there is an outdoor unhealthy air pollution alert. These hazards mainly come from three sources. One source is radon—an odorless, colorless gas which is a natural form of radiation resulting from a breakdown of uranium in soil, rocks, and water. Secondly, polluted air comes from changes in building and home furnishing materials—the substitution of plywood, particle board construction containing formaldehyde, and plastics for wood. Third is the growing use of chemicals in cosmetics, aerosol sprays, insecticides, and cleaning products in homes.

In the wake of the oil crisis in the early 1970s, the federal government issued strict energy efficiency standards for new homes, and federal tax credits and soaring fuel costs provided the incentive for millions of owners of older homes to insulate their houses. Unfortunately, a "tighter" home means restricted air circulation leading to problems similar to those found in office buildings. In older homes, all indoor air is exchanged with fresh outdoor air about once an hour. In extremely air-tight houses, complete replacement can take as long as ten hours, resulting in large buildups of potentially harmful airborne substances.

As early as 1984, the severity of the problem was becoming clear. The Chemical Hazards Office of the Consumer Product Safety Commission surveyed 40 representative homes in Oak Ridge, Tennessee and discovered between 20 and 150 chemicals in the indoor air depending on the time of day and year. Among the major pollutants found were carbon dioxide, carbon monoxide, sulfur dioxide, solvents, asbestos, plastics, pesticides, chloroform, and benzene. A variety of effects on the body may be caused by these chemicals, alone or in combination, including lung abnormalities, allergic reactions and possible reproductive effects.

You can reduce indoor air pollution in your home by some simple measures – mainly by substituting less toxic products. It is true that reading labels can be time-consuming and a nuisance. Taking the time, however, to read labels of products and substituting ones with natural ingredients and fibers is one of the few things easily within your power to do. Consumer groups have fought hard battles to gain legislation mandating labels providing information about the ingredients and composition of consumer goods so that you could be aware of what you were getting for your money. Now you can use this source of information to help you have a healthy pregnancy. With regard to household cleaning, it is remarkable what soap and water, baking powder, or vinegar solutions can accomplish. It is not always necessary to use commercial cleaners and detergents, most of which carry a *Keep Out of Reach of Children* warning. More about this later in the chapter.

In addition, be wary of synthetics. Try to purchase wood rather than plywood or plastic furniture. If cost is a problem, buy second-hand wood furniture rather than new synthetic substitutes. Buy natural fabrics like cotton and wool. Many new fabrics and carpets are made from petroleum and other chemical bases which have been largely responsible for the black toxic smoke that killed so many people in recent hotel fires.

Be careful of carbon monoxide which can seep into the house through cracks in a wall when the motor of a car is kept running in a closed attached garage or from woodburning fireplaces when the chimney is partially clogged. Carbon monoxide is also one of the byproducts of cigarette smoke, and if several smokers are present, cigarette smoke can raise the level of carbon monoxide above the acceptable level even in a room which is adequately ventilated for most purposes. Not only can carbon monoxide cause nausea, dizziness, and even death, but recent animal research has shown that carbon monoxide is harmful to the fetus. Carbon monoxide can cross the placenta. By doing so, it lowers the amount of oxygen that the fetus can obtain from its mother and hinders the ability of the hemoglobin to absorb and release the oxygen it normally distributes.

Radon

One of the most harmful of the indoor air pollutants is radon gas, an environmental source of radiation that is linked to lung cancer. Only in the last decade have scientists become concerned about the possible effects of radon buildup in private homes, even though the threat from exposure to radon gas presumably began when humans first began living in caves.

Originally, the radon problem was thought to be localized in areas where the soil contained radium or uranium such as the Reading Prong, a geological formation that stretches through parts of Pennsylvania, New Jersey and New York. Now it is known that the problem is more geographically widespread, as soil permeability may be even a greater factor than the uranium content.

Radon is measured in picacuries per liter (pci/l) and the EPA suggests that radon levels in houses be under 4.0 pci/l. As usual, experts disagree. The EPA estimates that 8 million of the country's 70 million houses surpass that level. Critics feel that the EPA overestimates both the level of danger and the extent. These radiation scientists believe only about 5 million homes exceed the 4.0 picocuriliter level and that remediation efforts should be concentrated on the approximately 700,000 homes having radon levels over the 10.0 picocuriliter level. If you are a smoker you should be even more concerned about the radon level in your home as some research indicates that any health risk from radon is increased by its combined effect with cigarette smoke – a "synergistic" relationship.

Women and children spend more time in their homes than men and are, therefore, likely to be more at risk from exposure to indoor air pollution. But don't panic as a result of media coverage. It is not clear how much of a hazard low levels of air pollutants are and the estimates of health effects of high levels are usually based on lifetime exposure. This is another area where you can do something about the situation. When you are pregnant, you want this information quickly. First determine whether or not you face substantial risks. Contact your local Department of Health, State Department of Health, or State Environmental Protection Department. Most of the Northeastern states have agencies and hotlines that provide information on how to check for radon yourself or can provide you with the names of experts in the field. These can be obtained from telephone directory assistance. The U.S. Environmental Protection Agency (EPA) has recently produced a booklet entitled *Radon Reduction Methods: A Homeowner's Guide*. This can be obtained from: U.S. EPA Public Information Center, 820 Quincy Street N.W., Washington, DC 20001 (1-800-828-4445).

Many times, moderate radon levels can be reduced easily and relatively inexpensively by sealing cracks and crevices in the basement, floors, and walls, covering open sump holes, or keeping some windows open. Higher levels will require more expensive and sophisticated techniques such as sub-slab ventilation techniques and air-to-air heat exchangers. Many states provide lists of approved radon removal contractors who can explain the various procedures and their costs. These companies bid on radon removal jobs just as contractors do on building jobs.

Household Cleaning Supplies

We worry a great deal about what might hurt our unborn children. A large weight is removed when we learn that our babies are born healthy. Yet we often pay too little attention to what might hurt our children after they are born. Most often, we are careless because we do not know that many common cleaning supplies can be toxic, especially to young children. An old fashioned punishment for children who used curse words was to wash out their mouths with soap. They survived to tell the tales, though few would support such a punishment today. Now we wash with detergents and think of detergents as being as benign as soap. No such luck! Detergents contain such chemicals as linear alkylate sulfonate and ethyl alcohol and dishwashing powders contain sodium metasilicate which causes burns. Thousands of children each year are poisoned by drinking household products. More are poisoned from dishwashing and washing machine detergents than from anything else. Perhaps this is because they are packaged in brightly colored plastic bottles and boxes and are left in easily reachable locations. We can reduce this hazard by keeping our cleaning supplies in locked closets or by simply using six inexpensive and effective substances that our grandmothers used: soap, vinegar, baking soda, ammonia, washing soda, and borax. The following is a list of cleaning tasks and some less toxic cleaners that can be used to perform them:*

General Surface Cleaning

Vinegar and baking soda are the best for this purpose. Mix several tablespoons of vinegar in a pail of water. Use this solution to clean counters, stove tops and appliances. The vinegar quickly evaporates and leaves no spots. More importantly, hazards are virtually non-existent in the use of a vinegar solution. For greasy areas and cleaning coffee pots, chrome, and tiles, use concentrated baking soda.

**Source:* Adapted from material in *The Household Pollutant Guide*, Center for Science in the Public Interest, Anchor Books, Garden City, NY, 1978, pp. 187-189.

Scouring

Baking soda is also excellent for normal scouring of sinks, tubs and sticky surfaces. Use in the same way you would use a commercial scouring powder. The abrasives in the commercial products can scratch the surfaces and the chlorine bleaches in them can cause skin and eye allergies.

Cleaning Pots and Pans

Solutions of vinegar or ammonia and scouring with baking soda should take care of most stains. If food is baked in, soak overnight before trying to clean. Buy stainless steel, enamel-coated aluminum, or pyrex cookware. They are preferable to plain aluminum. Tiny amounts of aluminum, of questionable health value, dissolve in the process of cooking acidic foods.

Bleaching

Borax is the most effective bleach and can be substituted for chlorine bleach, which chemically attacks the fabric as well as the stain. Borax just whitens the clothes and can be used with all colors and fabrics. *Never* mix bleach with ammonia as the combination produces a dangerous acid gas.

Oven Cleaning

The first choice for cleaning an oven is baking soda; the second choice is ammonia which gives off irritating vapors and can be toxic in high concentrations. A water solution of baking soda will take off grease. Sprinkle dry baking soda on crusted spots and leave in place for 5 minutes. Then scrub off with a damp cloth. Very stubborn spots can sometimes be scraped off. An alternative method is to place a solution of 1/4 cup ammonia mixed with water in a non-aluminum dish in the closed oven overnight. You can wipe off the stains with a damp cloth the next morning. A self-cleaning oven eliminates the need for most cleaning. Most homemakers find it a blessing, but make sure that the kitchen is well-ventilated as the fumes emitted during the high-temperature cleaning can be quite irritating. It is wise to use the baking soda to eliminate the excess grease before using the self-cleaning mechanism to avoid these vapors.

Cleaning Drains

There is nothing more aggravating for the homemaker than when a drain or toilet blocks up. Care in preventing food, hair, and grease from clogging the drain in the first place is the first line of defense, but in most households this first line of defense crumbles. Slow drains can usually be opened by pouring hot water down them followed by 1/4 to 1/2 cup of washing soda. After waiting for 1 to 2 minutes, flush with hot water. If the drain is completely blocked, use a plunger or small plumber's snake followed by the slow drain treatment. If you are not successful call the plumber. Commercial drain cleaners are very bad for human health and the pipes. Do not use these dangerous commercial products. The odds are that if the above suggestions are not effective the commercial ones will not be either.

For more information on household toxics, call: Toxic Infoline, 1-800-648-NRDC (in New York State call 1-212-687-6862). Also useful is the *Michigan Household Hazardous Substance Handbook*, which for the price of $15 includes an update service to keep the 3-ring handbook current with the latest toxics recommendations. You can obtain a copy by writing the Ecology Center of Ann Arbor, 417 Detroit Street, Ann Arbor, MI 48104 (313) 761-3186. You can also write for the series of pamphlets, fact sheets, and brochures called environmental hazards in the home. The package costs $2 and can be obtained from: The Connecticut Fund for the Environment, 152 Temple Street, New Haven, CT 06510.

Pesticides

From the time they lived in houses, wore clothes, and stored food, our ancestors had to contend with animals and insects who wanted to share their quarters and possessions. They discovered many successful methods of getting rid of these pests. While these may be less efficient than the pesticides we now use, they are also less toxic to us and are not likely to generate pesticide immune bugs. Two of the most powerful "anti-bug weapons" our grandparents used were cleanliness and sunlight. Now that so many of us hold two full-time jobs, one outside the home and one inside the home, we have had to skimp on the outdoor airing of bedding and thorough closet cleaning, chores considered a must in our grandmothers' and great-grandmothers' days. The ritual of spring cleaning not only gave us a neater, cleaner, fresher-smelling house, it also eliminated pests who like to live in warm, humid, and dirty crevices, among minute cracker crumbs and spilled milk stains in old clothes not worn for years.

In her book *Bug Busters: Getting Rid of Household Pests Without Chemicals*, Bernice Lifton offers a seven-point strategy: Build them out; starve them out; clean them out; shake them up; trap them; barricade them; repel them. Most of us abhor cockroaches, ants, wasps, mosquitoes, fleas, lice, bedbugs, rodents, waterbugs, and spiders, and we want to get rid of them quickly and forever. We often use inappropriate methods to accomplish this. A massive overkill assault with toxic commercial pesticides, particularly those in aerosol cans where pesticide vapor can float in the air and settle on you as well as the insects, is not the answer. The recommended non-toxic methods are slower, but it is much more important to insure your own and your unborn child's health than to rid your house in one swoop of the pests you have already been living with for awhile.

Basically the first non-toxic steps are to prevent insects and rodents entry into your home by repairing cracks and crevices and to prevent the welcome once the pests enter by reducing dampness, throwing out clutter in attic, basement, and garage, cleaning closets, drawers, and kitchen cabinets, under kitchen sinks, and garbage cans, and moving and cleaning under heavy furniture. Most importantly, all staples and cereals should be kept in tight containers.

Pesticides once thought of as saviors are now considered as possible carcinogens and reproductive hazards. Many of the pesticides used in agriculture have already been removed from the market because of their toxicity (DDT, Kepone, 2,4,5-T and DBCP) and more are suspect.

Fourteen thousand kinds of pesticides are currently used in the United States. Chemically they fit in five categories: chlorinated hydrocarbons, organophosphates, carbamates, botanicals, and inorganic compounds. Each chemical group contains some extremely toxic pesticides and some that are fairly safe for humans if used with caution. Be sure to check the ingredients in any pesticide you use since many pesticides have been associated with harmful reproductive effects in both men and women. The National Pesticide Telecommunication Network will provide you with additional information. Their toll-free number is open 24 hours a day, 7 days a week: 1-800-858-7378. All pesticides, no matter what their category, are somewhat toxic. The most important thing you can do is to read the label and only use the pesticide that mentions the specific pest you are trying to eliminate and only under the conditions the label describes. For example, do not use a product labelled "for outdoor use only" indoors just because you do not have anything else handy around and are phobic about insects or rodents.

We are protected to some extent by the warning labels on the containers of all poisonous substances. The problem is that we are not always sure exactly how to interpret the warnings:

> DANGER-POISON imprinted over a skull and crossbones means the substance is deadly and kills quickly, violently, and with very small doses. Do not keep any of these pesticides in your house especially if you have children.
>
> WARNING indicates a moderate degree of toxicity. The label may also say: harmful if swallowed. Use with extreme care.
>
> CAUTION indicates toxicity but to a lesser degree than those marked with the warning label. Use carefully.
>
> KEEP OUT OF REACH OF CHILDREN means exactly what it says. Little children display a great deal of curiosity and agility and manage to crawl, squeeze, and climb into places you thought were secure. If there are none of the other warning labels on the product, than the keep out of reach of children label indicates the least toxicity.

We must remember that all these cleaning agents and pesticides have a greater effect on children than on adults. Children under ten are the most frequent pesticide poisoning victims. (The section in

Chapter 4 regarding agricultural workers also provides information on pesticides.) Only use pesticides as a last resort. Use the least toxic possible. Instead of starting a long-term spraying campaign yourself, contact a well-trained, reliable professional to do the job quickly and effectively. Before hiring him, ask what chemicals he plans to use and check them with your State Department of Health or Environmental Protection Agency or call the National Pesticide Telecommunication Network.

Before you apply any pesticides, try a little patience first and some of the suggestions below. *The Bug Busters* book cited above provides many simple and practical suggestions in addition to the ones listed. It should be available from your local public library.

The next time you are tempted to reach for the spray can remember that ants are the natural enemy of the termite. Here are some less toxic ways to get rid of household pests.

1. Insects have a keen sense of smell and avoid sharp scents like mint, tansy, basil, and cedar. You can buy little jars, pierce holes in the lids, fill them with selected herbs, and place them in closets, drawers, and shelves.
2. Make simple traps and fill them with food harmless to humans but harmful to the pests. Meat or syrup can kill flies. Beer finishes off cockroaches. Silverfish die from eating flour. Also spring traps coated with peanut butter attract different types of rodents.
3. Spread technical (blue colored) boric acid bought in hardware stores (not to be confused with medicinal boric acid) behind the refrigerator, under the sink, and in corners to kill cockroaches. Vaseline barriers along the woodwork and countertop-wall edges might decimate an ant invasion as a temporary solution. For long-term control, trace the invasion back to the points of entry and fill the openings with petrolatum, putty, or plaster.

In addition to the dangers discussed so far, homemakers and household workers face other insidious hazards, such as noise from household appliances, injuries arising from moving and lifting heavy objects in order to clean under them, electric shocks from loose wires, radiation from leaking microwave ovens, tripping over toys the children have left out, and stress from long hours and heavy demands of housework and child care. Most of us can remember the discomfort of listening to sneakers rhythmically banging in the clothes dryer, wishing we could smash the machine, but knowing there was no alternative if we needed to dry them quickly.

*Dealing with Hazards in the Home**

Hazards

Noise—excessive noise from household appliances—washing machines, dishwashers, clothes dryers, vacuum cleaners—may lead to anxiety and stress.

Electric shock—from poorly maintained equipment.

Muscle and bone injury—performing repetitive tasks in uncomfortable positions, excessive standing, and lifting heavy objects.

Stress—long hours, low or no pay, and working in isolation can lead to increases in cholesterol levels, hypertension, headaches, frustration, and depression.

Accidents—burns from cooking, broken bones from slipping on wet floors and tripping over clutter.

Microwave ovens—continued exposure to leakage may be hazardous to eyes and reproductive system.

Formaldehyde—a carcinogen in animals which may also enhance the effects of other carcinogens and mutagens is found in home products, cigarette smoke, and indoor air pollution.

Drain cleaners—can cause burns and can produce toxic gases if combined with bleach or toilet bowl cleaners.

Oven cleaners—give off fumes that are irritating to breathe.

**Source:* Adapted from *The Household Pollutant Guide*, Center for Science in the Public Interest, Anchor Books, Garden City, NY 1978.

Ammonia and bleach – can cause eye and lung irritation and can form toxic gas if mixed with each other.

Cleaning fluids – may contain organic solvents which can cause irritations. Some of these substances can cause cancer.

Aerosol sprays – can irritate lungs.

Waxes and polish – can irritate lungs and nasal passages.

Pesticides – can cause serious poisoning.

Prevention

1. Substitute less toxic products.
2. Read labels and avoid using particularly strong chemicals.
3. Keep appliances and electric circuits in good condition.
4. Use gloves when handling potentially hazardous products.
5. Clean up clutter around the house.

THE NEED TO KNOW HOW TO EVALUATE RISKS AND HAZARDS

Protecting your own health during pregnancy, and the health of your unborn baby and your children requires continuous awareness of new information about risks and integrating the information into your daily life at home and in your paid employment. Many pregnant women read voraciously about the "do's and don'ts of being pregnant." What frustrates them is that they read so many conflicting statements and are presented with so many uncertainties. What good is it if scientists learn more about the occupational reproductive risks in the future when you are pregnant now? The next chapter, "The Information Gap: How to Judge Risks and Hazards," will explain why it is extremely difficult to accurately determine the harmful health effects of exposures to substances in the workplace and how to make judgments based on preliminary evidence. Even governmental regulating agencies do not always agree on what evidence to accept or what standards to promulgate. If they were to weigh pregnant workers needs as highly as they do companies' financial burdens involved in cleaning up the workplace, they would base their decisions on the most conservative estimates in order to provide the maximum protection. Pregnant workers are well aware that the "history of harm" has been one of a continual lowering of safe exposure standards; a level asserted to be harmless one year may be considered harmful a few years later. To achieve much needed changes in the regulatory agencies requires social action. Ways of going about this are discussed in the last chapter.

SUGGESTED READINGS

DeMatteo, B., 1985, *Terminal Shock – The Health Hazards of Video Display Terminals*, NC Press Limited, Toronto.

Liebman, S., 1982, *Do It at Your Desk: How to Feel Good from 9 to 5. An Office Workers Guide to Fitness and Health*, Tilden Press, Washington.

Michigan Household Hazardous Substances Handbook, 1986, and *Toxicants in Consumer Products*, Municipality of Seattle, WA, can be obtained from the Ecology Center of Ann Arbor, 417 Detroit St., Ann Arbor, MI 48104, (313) 761-3186.

9 to 5 Newsletter, published 5 times a year by 9 to 5, by the National Association of Working Women, 614 Superior Ave., NW, Cleveland, OH 44113, (216) 566-9308. All members receive the newsletter. Yearly membership is $15-25, based on your income. For non-members, subscriptions are $25 for individuals and $40 for institutions.

Stellman, J. and Henifin, M. S., 1989, *Office Work Can Be Dangerous To Your Health* (revised and updated edition), Fawcett Crest, NY.

U.S. Department of Labor, Women's Bureau, 1985, *Women and Office Automation: Issues for the Decade Ahead*, 200 Constitution Ave., N.W., Washington, DC 20210.

Wallace, D., 1982, *The Natural Formula Book for Home and Yard*, Rodale Press, Emmaus, PA.

PART III.

STRATEGIES FOR THE FUTURE

Photo © Earl Dotter

Chapter 6

The Information Gap: Learning to Judge Risks and Hazards

AT RISK OR NOT AT RISK, THAT IS THE QUESTION

The AIDS scare just came out while I was pregnant. Nobody knew what was happening or whether we could catch AIDS through the blood we were processing. They didn't know what to do because they didn't know what was causing it. They didn't have any testing for it. It was touchy for a while.

—Maria, lab technician in a blood bank

Many of my pregnant friends who were flight attendants worked the initial six months the airlines allowed, and some of them had children with handicaps. You always wondered if the handicaps came from their having stayed. You really have no way of knowing if the proportion was any greater than the proportion of people in the ground jobs. I have a very good friend whose child ended up with cerebral palsy. I had a few friends that lost their fetuses during those six months. So you don't know. Did the plane carry some dangerous chemicals in the cargo or was it due to airline conditions, mainly decompression?

—Nora, flight attendant

We live in an age of uncertainty that affects how we live our lives and how much control we feel we have over our lives. We have three choices: putting our heads in the sand and ignoring risks and hazards; refusing to make any judgments by using the rationalization that everything is harmful, so why bother; or attempting to judge risks and hazards in our communities, workplaces, and homes within this uncertainty. This is a very difficult task and few of us can do it well. We lack the necessary information, skills, and training. It is somewhat like trying to figure out what the completed jigsaw puzzle looks like when we only have a few pieces and no copy of the picture. Many of us do not have the patience to be environmental and occupational detectives. We give up in frustration, pretending that nothing bad will happen. Yet underneath the facade, we are worried and would like to reduce our risks.

Being a pregnant working woman adds an additional dimension to this task because you are concerned about the effect of occupational exposures on your unborn child's health as well as your own. How can you know which of your fears are well-founded and which are not? How much evidence is required before a decision is made that a substance harms reproductive health? How do regulatory agencies evaluate the low risk of a severe hazard as compared with a high risk of a moderate hazard when making rulings?

This chapter presents basic information about evaluating risks and hazards so that you can learn to make some sense of what you read and hear and can begin to understand why scientists give conflicting findings and interpretations. It also tries to dispel common myths such as "all chemicals cause cancer" or that chronic low-exposure is *always* as bad as one big exposure.

The information can be used by you and your partner if you are thinking about becoming pregnant or are pregnant and are worried about the safety of your workplace. The most important questions for you to ask are listed at the end of the chapter. You can make a copy and easily carry them with you if you decide to investigate further any possible reproductive workplace, home, or neighborhood hazard that worries you. (See the glossary at the back of the book for more detailed explanations of the scientific terms used.)

Although more and more pregnant women continue working in the paid labor force, little research has been conducted on reproductive harm from workplace hazards. Even scientists do not know to what extent many substances are hazardous. They disagree about the interpretation of existing evidence, but agree that much more data are needed before firm conclusions can be made. For example, a committee of the National Research Council reviewed the identification and testing of known and suspected toxic chemicals. Shockingly, they found that relatively few out of tens of thousands of important commercial chemicals had been tested extensively; most had hardly been examined, and only 27% of the 664 toxicity tests were judged acceptable scientifically. Even when a chemical is shown to be toxic, it is harder to pinpoint reproductive harm than it is to determine other health effects. One difficulty is that a substance causing harm in one species may not cause harm in another because of differences in the way the substance is processed in the body and differences in the reproductive systems of various species. For example, thalidomide, a drug taken by women during early pregnancy as an anti-nausea drug, caused their babies to be born without normal arms or legs. Thalidomide damages the fetuses of rabbits and non-human primates such as baboons, but not those of rats and mice, which are the animals used in most toxicity tests.

The time at which exposure occurs also may be crucial for reproductive damage. Some hazardous substances act on the development of the sperm (spermatogenesis) but not later, while others can damage developing ovaries but not mature ones. Furthermore, some toxic substances only damage one function of one particular cell while others are toxic to many sites within the human body. Another major problem is trying to establish whether there may be a safe upper limit of exposure to substances that can cause cancer, genetic damage, or developmental defects in humans.

Because understanding of reproductive hazards is clouded by such uncertainty at every level of investigation, the answers we need are not easy to obtain. If we want to continue working and protect our pregnancies, we have to understand the way risks and hazards are judged within the present limitations of scientific knowledge, and the roles scientists, consumers, businesses, and the government play in this process.

It is also important to understand how the mass media interpret and present risks to the public. Most of us get our information from newspapers, magazines, and T.V. We often assume these reports to be more accurate than they actually are. Scary headlines sell newspapers, and sources of advertising revenue color interpretations. Often an article cites a substance as causing a specific illness and a later article denies it.

> During this pregnancy I was worried about an article in the newspaper about the water supply in this area causing a high risk of cancer because the pipes have asbestos linings. The first thing I thought of was, "What kind of water am I drinking?" A week later another article said that people shouldn't get upset about the water as the scientists weren't sure whether the water or something else was causing the increased risk.
>
> —Clarissa, operating room nurse

As Clarissa's experience indicates, media reports should be viewed as "alerts" to search for further information not as "facts" to be acted upon.

MEDIA MESSAGES

There are certain biases in media reporting. Once we determine what they are, we can discount some of what we read and concentrate on deciding what further questions need to be answered for us to make an informed decision. Newspapers, magazines, and television need to attract large audiences in order to survive financially. Increasing circulation and ratings are the name of the game. The media over-report dramatic events involving a large number of injuries and fatalities that occur at one time because these stories attract readers and viewers. The more common everyday risks and hazards causing an equivalent or even larger number of diseases, injuries, or premature deaths over a longer period of time do not get equal coverage. These occupational and environmental exposures may be of more concern to society, but they are not news.

Reporters write under deadlines. It is often easier for them to obtain information from government offices or corporations that have community relations departments than to track down the individuals exposed to toxic substances and obtain their side of the story. Valuable information contradicting the highly publicized version can only be found in newsletters mailed to members of public interest and women's groups. Mainstream newspapers and magazines appear on every newsstand; the advocacy press is harder to find. Check with the main newsdealer in your community and library about these consumer-oriented, public interest publications. These have their own biases, but they are usually biased in favor of the consumer. It is wise to read articles about a hazard that you are worried about in a wide range of publications. You will learn how much agreement there is in interpretation of the risk involved and the reasoning behind the differences of opinion (See Figure 6.1.)

The media also take editorial positions. They feature stories dealing with those social issues and perspectives they deem important. Both sides of issues are usually presented, but one side gets more coverage and is highlighted. Women reading about reproductive hazards must be aware of these biases when they are sifting and evaluating what they read. They must ask how complete and accurate the reporting of the risks is and what sources were used. Versions and biases differ within the mainstream press and between the

FIGURE 6.1. Some Mainstream and Advocacy Sources of Information

Mainstream Press	Advocacy Press
Chicago Sun	Citizen's Clearing House for Hazardous Waste Action
Los Angeles Times	Ecological Law Report
Newsweek	In These Times
New York Times	Mother Jones
Philadelphia Inquirer	Natural Resources Defense Council Newsline
St. Louis Post Dispatch	National Women's Health Network's Network News
Time	Public Citizen Newsletter
U.S. News	
Washington Post	

mainstream and advocacy press. Here are some questions to ask when you are reading an article on workplace exposure:

1. Who was the source of the information?
   a. Government agencies
   b. Businesses and corporations
   c. Labor
   d. Academic scientists
   e. Public interest groups
   f. Women's groups
2. Was the information checked with other sources?
3. Was more than one perspective presented?
4. How many studies was the article based on?
5. Were the studies conducted on humans or animals?
6. Does the article tell you who funded the study?
7. Does the article tell you about the weaknesses of the study and whether more research is needed before the results should be acted upon?
8. Does the article tell you where you can obtain further information about the scientific report?

If most of these questions are not answered in the article, you should be skeptical, at least initially, about its accuracy and impartiality because it gives you no evidence of factual information upon which the article was based or of possible biases in the reporting. More complete coverage will include specific details and names of specific scientists, universities, corporations, or government agencies involved in the research. You can then follow these leads to gather the additional facts you need to come to a decision about the workplace reproductive risks you face.

HOW DO SCIENTISTS DECIDE WHAT IS DANGEROUS?

Scientists generally have to base their decisions on rough estimates derived from human, animal, and bacterial studies, because there is no firm proof that a certain amount of exposure to a specific toxic substance results in a specific degree of harm to our reproductive health or the health of our fetuses. Actually, very little is known about substances in the workplace that cause changes in human genetic material (mutagens) or affect the development of the fetus (teratogens). More is known about cancer-causing substances (carcinogens). Even in this area, though, most of the decisions concerning human risks are based on scientifically-based estimates derived from many different kinds of animal investigations. Similarly, animal studies are used to determine reproductive damage.

Occasionally substances are found to cause cancer because humans have been exposed to very high doses of them or lower doses for very long periods. This has happened with asbestos workers, uranium miners, and soldiers participating in atomic testing. Researchers refer to these kinds of investigations as high-dose epidemiological studies. High-dose refers to the amount of exposure these individuals received. Epidemiological studies are those which investigate the effect of a suspected toxic substance on a group of people.

Studies of high-dose exposure have severe limitations, though. Even after the epidemiological study is completed, scientists still may not know whether a particular substance is dangerous at a low dose or for a short period, if so, how dangerous it is, or at what level it becomes dangerous. Ideally, researchers like to have human evidence, but this is often hard to obtain. Either scientists have to study thousands of people at one time or a smaller group of people over their lifetimes. The former is almost impossible to do, and the latter takes too long. Most risk estimates, therefore, are based on animal or bacterial studies.

It is especially important for you to understand what is meant by scientific and statistical evidence. Often this material is difficult to understand and research scientists seldom possess the expository skills needed to present the issues simply and clearly. Too often you need to read the explanation several times before it makes sense. Management also has difficulty evaluating risks and hazards. When concerned about reproductive hazards, they focus on pregnant women. Therefore, you need this knowledge not only to evaluate general health hazards to yourself and your unborn children, but to defend yourself against job discrimination masquerading as protection (Fetal Protection Policies). Too often management will remove you from your job rather than clean up the workplace for everyone.

ANIMAL STUDIES: RATS AND MICE IN THE LABORATORY

Most of us dislike rodents and view them only as pests to be exterminated, but they have played a major role in the scientific fight against human disease, because a large number of animal experiments are conducted on mice or rats. The studies compare rats and mice in experimental groups with those in a control group. Many more animal studies have been conducted to determine whether exposure to specific chemicals cause cancer than have been conducted regarding reproductive harm.

Cancer Studies

In cancer studies, the mice or rats in the experimental group are exposed to specific chemical doses during their lifetimes of 24 to 36 months. Generally the lowest dose used bears some relationship to anticipated human exposures while the highest dose may be hundreds or thousands of times higher. Another group of mice or rats act as controls. This means that they are treated exactly the same as the animals in the experimental group but not exposed at all to the chemical being studied. Scientists usually call this a zero dose. At the end of a cancer study, the scientists count the number of tumors or lesions that grew in the animals receiving the different doses of the chemical substance and compare these results with the number of tumors found in the control group. For example, if 10% of the rats in the control group develop tumors while 50% of the rats in the experimental groups develop tumors, the researchers would be fairly certain that there were toxic effects because the difference was so large. If, however, 10% of the rats in the control group developed tumors and 13% of the rats in the experimental group developed tumors, it would be harder to pin the blame on the toxic substance being investigated.

Cancer studies take several years and are very expensive. The care and feeding of the experimental and control groups of rodents is a tremendous task and may rival childrearing in sleepless nights, and unexpected hazards. Sometimes the investigation requires that 700 rats be fed and tended and as many as 500 of them receive measured doses of substances every day for two years. Lab workers must record the general condition of the rats every day and weigh them twice a week. Every rat that dies before the experiment is concluded must be examined and have its death recorded, and samples of its tissue must be stored for future analysis.

At the end of the study, each surviving animal is killed; the main organs are analyzed and stored in formalin, and tissue must be prepared for microscopic examination. A typical 2-year study can generate nearly a million pieces of information plus 250,000 slides.

With this amount of work to complete, lab technicians can make mistakes in many different ways. Before international agreements on Good Laboratory Practice Standards went into effect, occasional horror stories emerged. Water bottles remained empty for days because the workers didn't come in on weekends and nights when they were supposed to, or rats became sick and died because lab workers were too busy to clean the cages properly, allowing bacteria to multiply.

How accurate are the records produced by these studies? The recently instituted Good Laboratory Practice Standards include inspections, so the previous laxity has been for the most part eliminated. Many chemical substances still on the market, however, are deemed to be safe on the basis of studies conducted before the standards went into effect—if they have been tested at all.

Drawing conclusions about humans from test results on animals (extrapolation) is especially difficult because some of the animals routinely develop tumors that are not caused by the substance being tested. Scientists face two main problems in applying findings from animal experiments to humans. One is that susceptibility among animals or between animals and humans differs. Certain chemicals give rats cancer but do not seem to harm mice or vice-versa. Not only do different species react differently to a specific substance, but individual members of a given species also can have different reactions. Will they harm humans? Nobody knows for sure on the basis of these animal studies. The second problem is to determine how accurately findings based on high doses of toxic chemicals given to animals over a short period of time reflect the risk humans face from the lower exposure levels they experience in their environments over a longer period of time.

Reproductive Studies

Reproductive studies are usually conducted over a much shorter time span than cancer studies. In a reproductive study, the investigators concentrate on reproductive markers such as birth defects or still births in the litters, or pregnancy rates of the females in the experimental groups, compared with those in the control group. The short pregnancy of 22 to 23 days is one of the reasons that mice and rats are particularly suited to be experimental animals for reproductive studies. (Many of us feeling the discomfort of pregnancy would welcome a 22-day gestation period from conception to birth, or even a 22-week one.)

These studies are generally of two types. First there is a general reproductive and fertility test which investigates the effect of a suspected toxic substance on fertility, pregnancy, and its transfer through the mother's milk. The second kind is called a teratology test, specifically designed to determine the effect on fetal development.

General Reproductive-Fertility Studies

In general reproductive-fertility studies the mice or rats are usually divided into 3 experimental groups, each receiving a different dose of the substance, and 1 control group. The substance can be injected, rubbed on the skin, or combined with food. The control group either receives nothing or a placebo (an inert substance).

The treatment of males starts 60 to 70 days before mating. This is their sperm cycle—the length of time that sperm take to mature fully from stem cells to "adult" sperm cells capable of uniting with the egg and forming an embryo.

The female mice or rats are treated with the decided upon doses of toxic material 14 days before mating. This covers 3 to 4 ovulation cycles. At the end of the 14 days, the treated females are mated with the treated males. Pregnancy begins on day 1 when the egg is fertilized; implantation (when the fertilized egg embeds in the womb) occurs on day 6. The period when the organs and structure develop (embryogenesis) occurs from day 6 to day 15. Birth occurs at 22 to 23 days. The offspring are "nursed" by the mother during the 3-week period of lactation, at the end of which they are weaned.

As we know, infertility is a fact of life for too many young couples. Controlled experiments ask whether chemical substances can be blamed by comparing the number of successful matings within each treatment group and the number with the control group. If scientists want to know whether the fertility of the male or of the female is affected, they can mate 1 male with 2 females, one from the experimental group and one from the control group. If neither female becomes pregnant, the problem probably lies with the male, but if only 1 female becomes pregnant then the toxic is likely to have affected the female. If both become pregnant, fertility has not been influenced by the treatment.

Miscarriages and breast milk pollution are two additional worries of pregnant workers. Again, controlled animal experiments can help shed light on these aspects of reproductive health as well. If the female delivers a normal-sized litter and the offspring stay alive till weaning, you have a measure of the mother's ability to carry to term and to feed her young successfully. If the baby mice and rats start dying off during the nursing period, then toxic effects are most likely being transmitted through the mother's milk. The remaining offspring can also be killed at weaning and autopsied in order to ascertain whether any further damage has been done. (See Figure 6.2.)

It would seem that malformations could also be detected by these reproductive-fertility experiments, but this is not the case. The mother rodent usually eats a defective offspring immediately after its birth, and births often occur at night when lab workers are not in

FIGURE 6.2. General Reproduction and Fertility Test Using Rats or Mice

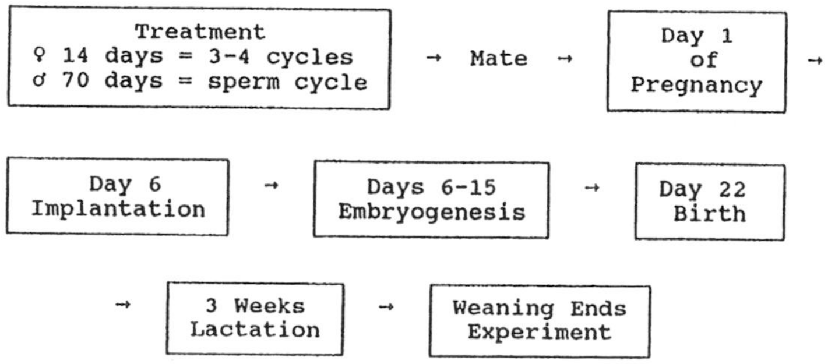

attendance. Therefore, another type of experiment called the teratology test has been designed. In this kind of experiment, neither the male nor the female in any of the treatment groups is exposed to the toxic substance before mating or during the period of pregnancy before implantation.

Teratology Test

The different doses of the toxic treatments are started for each group during the developmental period (embryogenesis), from day 6 until day 15. On day 21 (the day before expected birth), the pregnant rat or mouse is killed and the fetuses analyzed. While this is hard on mother rat, much worthwhile data can be gathered. Because the dead mother cannot eat the malformed fetuses, the researcher can count how many there are and what kinds of defects they have. The researcher can also look for little blobs of tissue that indicate a reabsorption—equivalent to a spontaneous abortion in humans. If the embryo or young fetus dies in the uterus, it is reabsorbed rather than expelled. Stillborns can also be counted. These are fetuses that died too close to the due date to be reabsorbed. Finally, each live fetus can be weighed and the weights of the fetuses are compared to the dose level of the toxic substance received by the mother. These four measures—malformations, reabsorptions, stillbirths (dead fetuses), and body weight—are compared between all of the three experimental groups and the control group. (See Figure 6.3.)

What can we learn from such experiments? A great deal—but as is true with cancer studies, whether the findings apply to humans is not always clear. In general scientists feel more certain that these

FIGURE 6.3. Teratology Test

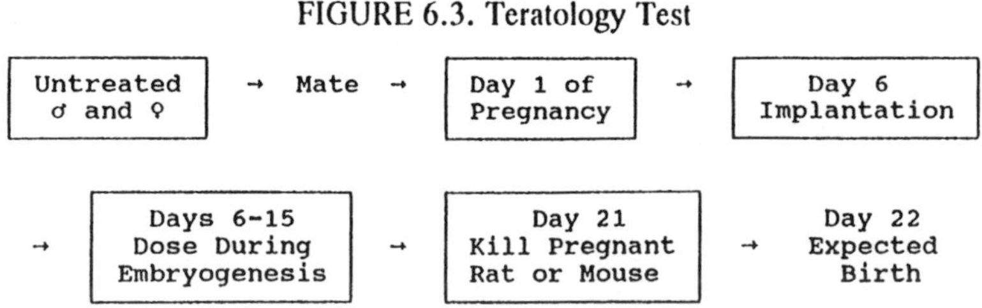

results would apply to humans if the same results were obtained with two or more kinds of animals, e.g., rabbits and mice.

BACTERIAL STUDIES: MILLIONS OF MICROBES UNDER THE MICROSCOPE

Considering the large backlog of substances to be tested, plus the 1000 new chemicals entering the market each year, some experts feel that animal testing is too time-consuming, too slow, too costly, and too uncertain to continue to play the key part in the regulatory process. They see a major role for new tests using bacteria.

Several of these new bacterial tests can be completed in a month and cost around $5,000, whereas animal tests investigating effects of workplace toxics on cancer can cost more than $600,000 and can take from 4 to 7 years to complete. By measuring how toxic chemicals affect genes, chromosomes, and other cellular systems, scientists may be able to predict more exactly the substance's capability of causing cancer, mutations, birth defects, or environmental damage.

Scientists have been slow to study male and female workplace-related reproductive hazards, but new tests using living bacterial or animal cells instead of live animals have stimulated investigations in this area. It is hard to imagine that a lab technician can examine a billion bacteria in a 3-square-inch area, but that is what is happening. Many scientists are excited about these types of studies because such large numbers of living organisms can be examined so quickly. Genetic material is similar in all living organisms and millions of bacteria can be studied at one time. These large numbers are needed in order to find evidence of mutations, which are fairly rare and which could trigger cancers that would normally take many years to develop in whole animals.

The development of cellular tests has excited many researchers. Others, however, are more cautious. They believe that while these tests are quicker and cheaper, they are not necessarily better predictors of human toxicity. Bacterial studies can only test for genotoxics (substances affecting genetic material in the cells), but these are the key ones and many scientists believe that if they could identify, eliminate, or at least reduce the use of the important genotoxics, they could go a long way in cleaning up the workplace. Genotoxics usually attack body (somatic) cells rather than germ (ova and sperm) cells, which are better protected. If, however, these genotoxics do attack the germ cells, malformations in the offspring could result.

By comparing results of the animal and bacteria tests, scientists will be better able to decide how well these shorter, cheaper tests predict potential health hazards of chemicals. So far cellular tests by themselves are somewhat unreliable and may overpredict hazards, so animal testing with all its drawbacks still remains the best, though far from adequate, way of determining long-term chemical hazards.

HUMAN STUDIES: THE FINAL PROOF?

After the evidence from animal and bacterial studies is accumulated, scientists then have to determine whether humans are going to experience similar effects. It is interesting to learn about what happens to the offspring of other living creatures, but we want to know whether our own babies will be all right. To help answer these questions, researchers conduct epidemiological studies on fertility of male and female workers and pregnancy outcome. In some investigations, they compare the reproductive health of groups of workers who were exposed to certain substances with those who were not, or compare work histories of pregnant workers who suffered ill effects with those who did not to see what might have caused the harm. These epidemiological studies are primarily of two kinds—*retrospective* and *prospective*.

Take the case of researchers who are investigating the relationship between miscarriages and exposure to anesthetic gases that leak into the operating room. In a retrospective study, the researchers would compare nurses who had miscarriages with nurses who didn't have miscarriages, making allowances for other possible factors that might account for the outcome, such as age, number of previous children, history of miscarriages in the family, number of years employed, possible exposure to other hazardous material, etc. They would see whether a higher percentage of nurses who miscarried worked in the operating room than worked in other areas of the hospital.

In a prospective study, researchers would compare the reproductive patterns of a matched group of operating room and non-operating room nurses over a number of years to determine whether the miscarriage rate was higher among the operating room nurses than among nurses working elsewhere in the hospital.

When you read about a research study, you should ask the following questions:

1. Was the research conducted on animals, bacteria, or humans?
2. If it was conducted on animals, was it conducted on more than one kind of animal (e.g., monkey and rabbit) and were there harmful effects in more than 1 species?
3. How frequently do these kinds of mutations, birth defects, or miscarriages occur in the general population (the background rate)? Sometimes the background rate is not known accurately. As a rough guide, you can use the figure of 1 in 5 to 1 in 10 recognized spontaneous abortions and 2.5% of infants having malformations at birth.
4. Is the sample size used in the study large enough to show whether there is a significant increase (according to scientific criteria) due to exposure to the toxic substance?
5. Were other factors that might be associated with the harmful effect controlled, i.e., sex, age, race, education, occupation, life-style, residence?
6. What was the design of the study? Were people who were exposed to a suspected toxic substance examined and then followed for several years to see what ill effects they suffered? Were people who suffered from some reproductive harm examined and then their past investigated to determine if there was something in common they were exposed to?
7. How harmful is the substance? For example, scientists have found that saccharin, the sugar substitute, and DDT, the pesticide, cause tumors in some animals. It takes, however, far larger quantities of saccharin than DDT to produce equivalent

harmful effects. Furthermore, does any amount of exposure cause harm or is there a threshold below which there are probably no ill effects?

8. If exposure to the pregnant worker was being studied, did the article discuss any research investigating reproductive harm to men from exposure to the same substance?

9. What kinds of harmful health effects occur? There are four general types. (1) harm to developing fetuses (teratological effects); (2) changes in genetic material (mutational effects); (3) cancer causing; (4) damage to specific organs such as the lungs, liver, kidney, or brain.

Clinical and autopsy studies also provide valuable information. In clinical studies, scientists look at human (or animal) blood or bone marrow cells under the microscope to detect abnormal chromosomes, a possible indication of exposure to a mutagen (cytogenetic studies). Autopsies are very useful because they may identify the specific cause of death. They are rarely required, however, and coroners are not trained to look for possible occupational causes. Furthermore, the idea of an autopsy upsets many of the families. They refuse to give permission to have their loved ones "cut up." Valuable information, then, is never collected.

Company doctors are another source of vital information that is seldom obtained. If workers in one type of industrial plant become ill or die from a particular disease more often than the population at large, it is likely that some substance or combination of substances is responsible. Company doctors usually have access to this data but seldom speak publicly about it.

We are only beginning to link occupational exposure of both parents to sterility, miscarriages, birth defects and childhood cancer. Even when the focus is only on the mother, much valuable information is not collected. In many states, neither the occupation of the mother nor the father is recorded on the birth certificate of a newborn baby or death certificate of a stillborn or young infant. In fact, some states formerly listed the father's occupation but have removed this question. Parents view the birth certificate as a personal legal document and some resent answering existing questions on race, education, and ethnic background, considering them governmental invasion of their privacy. Fear of government prying has made it more difficult to trace the effect of parental occupational exposure on the offspring.

In order to obtain badly needed information, several states have instituted birth defects registries in which detailed parental and environmental information is collected. Birth defects data bases have also been developed in many industrial countries and some have been linked to computer networks. Medical researchers share information and report their findings at international forums. It is hoped that these worldwide efforts will hasten understanding of the causes. (Table 6.1 at the end of this chapter lists substances currently thought to cause adverse reproductive health effects in animals or humans due to occupational exposure.)

THE AGENT ORANGE STUDY

The specific cause, or combined causes of harm to reproductive health are difficult to pinpoint and isolate. If scientists do not have a standard of comparison, they cannot be certain that an increase in miscarriages, birth defects, or infertility has actually occurred or to what particular occupational exposure an increase might be attributed. The best way to determine if there is a cause and effect relationship between an exposure and an illness or defect is to study large numbers of exposed people and examine their health compared to an equivalent group of people who were not exposed to the hazard. Let us see what is involved in carrying out such a study and why it may leave many questions still unanswered.

The Agent Orange study of Vietnam veterans exposed to defoliants conducted in the early 1980s has received a great deal of public attention. Vietnam War veterans who thought they were exposed to Agent Orange, a pesticide that has been shown to have toxic effects in animals, were extremely concerned about the possibility of their increased risk for fathering babies born with birth defects.

In response to political and scientific concerns, researchers designed a large scale study of the possible linkage between exposure to Agent Orange and birth defects. Between 1968 and 1980, researchers identified over 7,000 babies in the Atlanta, Georgia metropolitan area as having severe birth defects. They also included a sample of babies born at the same time without known birth defects. This is called a case-control study. Information about the families of babies in the case group (those with birth defects) and those in the control group (those without birth defects) was gathered. The scientists placed special emphasis on obtaining a history of the father's military service, whether he served in Vietnam, whether he was likely to have been exposed to Agent Orange, and if so, for how long. The veterans' self-reports and an exposure opportunity index (EOI) score graded by the army Agent Orange task force were used to determine the level of paternal exposure. Both of these measures contain some inaccuracies but were the best that were available at the time.

The study concluded that Vietnam veterans who had greater estimated opportunities for Agent Orange exposure did not seem to be at greater risk for fathering babies with all types of defects combined. The researchers, however, did find these veterans had slightly higher estimated risks for certain types of defects. These scientists, however, could not determine whether these seemingly higher risks could be chance events, the result of some experience in Vietnam, or some other factor not yet identified. Other types of possible reproductive hazards harm were not examined in this study because the birth defects registry used to obtain the lists of babies born with severe defects does not collect information on effects such as infertility, miscarriages, or physical and mental handicaps that become evident later in childhood.

The ability of a study to identify increased risks for fathering a baby with birth defects depends on several factors.

1. *The size of the true risk.* If the risk is extremely high, it is more likely that a study will find this evidence than when the risk is much smaller. For example, if the risk of injury from being exposed to a harmful substance is ten times that of individuals who were not exposed, the effects are easier to detect than when the risk of exposed people is only 1 1/2 times as high.

2. *The number of cases available and the number of controls available.* If the harmful effect occurs very infrequently in both the exposed and non-exposed population, then the researcher would

need a very large number of individuals to investigate and/or would need to observe them over a long period of time.

3. *The rate of exposure being studied.* In the Agent Orange study, the case group consisted of babies born with severe birth defects and the control group consisted of apparently healthy babies. Thus the two groups were divided by the outcome of pregnancies. What the scientists are looking for is whether there is any connection between the exposure of the father and the rate and type of birth defects seen in their children. When the healthy babies were chosen from the state vital records unit, they did not know how many of the fathers had either served in Vietnam or had been exposed to Agent Orange. Based on data obtained by the Veterans Administration, it was likely to be between 10 and 20%. If the number had been very small, such as 1 to 2%, then there might have been too few fathers exposed to Agent Orange to be able to judge whether the exposure had an adverse impact.

WHAT MAKES SOMETHING A HAZARD AND HOW DO YOU MEASURE IT?

Before we can talk about hazards and risks we must define the terms. Usually, the term "hazard" refers to the substance or condition that is suspected of causing harm and the term "risk" for the probability of harm arising from the individual's exposure to that substance or condition. For example, lead is a workplace hazard for a pregnant woman. The risk is the likelihood, or probability, of her having a miscarriage, or giving birth to a child with a birth defect, as a result of her particular exposure. These are the definitions used in this book. However, in many articles and media accounts, these two terms are used interchangeably. For example, the term risk is sometimes loosely used to refer to harmful outcomes such as cancer, miscarriages, or sterility.

The way it works is that scientists first identify the hazard. Then they identify the relationship between exposure to the hazard and its unhealthy effect on humans. After that they try to determine the number of people suffering from this unhealthy effect and whether certain groups may be at greater risk. If exposure to X rays is at issue, researchers would be particularly interested in investigating the effects on patients receiving high doses of X rays and workers in occupations involving the use of X rays. Exposure is an important aspect to consider because many chemicals only exert harmful effects at high exposures.

Once a hazard is identified and scientists have studied its impact, the regulatory agencies have to decide whether to ban that substance entirely or determine an acceptable risk level. With regard to cancer-causing agents, the regulatory agency usually chooses as an acceptable risk level a dose that gives a probability of having one animal in 100,000 develop a tumor over a lifetime. This is roughly the equivalent of 30 tumors per year in the United States population, assuming that the entire population was exposed to the same dosage as the animal and assuming that humans have the same sensitivity to the chemical as the test animals. This is then called a "virtually safe dose" (VSD). In real life, of course, some of you are more exposed than others and some of you are more vulnerable than others. Furthermore, these VSDs are based on animal data, and as we saw earlier in this chapter, animals are different from humans. Finally, these VSDs are estimates for cancer, not for reproductive harm, so a VSD can be used only as a crude guideline in estimating a risk for an individual in your particular situation.

Part of the scientific argument about health effects also revolves around the question of whether there is a level of exposure below which there is no harm, and if so what that level is. For example, very low doses of dangerous pesticides used to be considered safe. Now this is controversial. Due to limitations of both animal data and mathematical models, experts do not universally agree upon the answers.

In light of scientific uncertainty, what should government regulatory agencies do? One possibility is that the government could outlaw any substance that posed a risk of cancer to any animal, regardless of the size of the risk. Congress passed a law called the Delaney Clause which forbids the use in foods of any substance that appeared to cause cancer in animal experiments. For example, under this clause some of the food additives such as specific red dyes used to color maraschino cherries were banned.

Where animal studies showed the risk of cancer to be extremely high, regulatory agencies also banned the use of non-food substances like DDT and chemicals whose industrial processes generated Dioxin as a byproduct. But it is impossible to ban all suspected carcinogens, mutagens, and teratogens (substances that harm the developing fetus). Therefore, scientists return to the concept of a virtually safe dose (VSD).

Unfortunately, the estimates of a VSD depends not only on the data gained from the animal studies but on the mathematical model chosen for translating the animal data into information that can be applied to humans. This means that if scientists use different mathematical models to interpret the same data, they will arrive at *different* VSDs which may be far apart. Faced with these diverse estimations, researchers concerned about the long-term as well as immediate health effects on workers support use of the mathematical model that estimates the VSD to be a very low dose. They choose this estimate even if a higher dose might not be harmful because they would rather err on the side of safety, believing that the best public health decision is to offer the greatest possible protection to the people exposed. This is especially important as current scientific thought suggests that carcinogens do not have a threshold (an upper limit below which no cancers are produced).

Federal regulatory agencies are now attempting to make risk assessment judgments more consistent and precise. Despite the lack of precision and scientific agreement, the regulatory agencies argue that risk assessment is the best tool they have for deciding priorities among environmental and occupational health problems.

Even when there is agreement on risk assessment, there may be no agreement on what is an acceptable level of risk for individuals to face. How many individuals have to suffer ill effects before government action is taken to ban or limit the substance? Numerous suggestions for determining acceptable risk levels have been proposed. Some are based on a standard of relative risk rather than absolute risk and recommend that risks from comparable activities should be used as a bench mark for acceptable risk. In Great Britain, Lord Rothschild proposed that a risk of exposure to a cancer producing substance should be considered unacceptable when the chances of death are above 1 per 7,500 a year, the risk of being killed in a car accident in Great Britain. A better idea may be to make both automobile driving and the workplace safer.

CONSUMER VERSUS SCIENTIFIC, BUSINESS, AND GOVERNMENTAL VIEWS OF RISKS AND HAZARDS

You will probably have different views about risks and hazards than scientists, businessmen or members of governmental regulating agencies. Pregnant workers see risks in terms of their lives and the lives of their unborn children that will be damaged by toxic substances in the workplace. They do not want to be the guinea pigs if insufficient testing is done before a chemical is allowed on the market. Even more importantly, they are concerned about the thousands of chemicals already available that either have not been tested at all or inadequately tested.

Risk-taking choices are based on women's values and not on some rational calculations of costs and benefits. They are shaped by their own experiences with risk-taking, other people's experiences, and those they hear about on television or read about in magazines and newspapers. Much information about environmental and occupational hazards comes from articles in newspapers, magazines, and TV programs. Thus, the media is likely to be important in shaping attitudes and behavior.

Scientists, on the other hand, strive to maintain the integrity of their professional disciplines and require a high standard of proof. They want to be 95% or 99% sure that their conclusion is not due to chance alone before they will state that a given substance does or does not cause harm. (This is technically known as statistical significance.) Obtaining this high level of proof depends on: the number of individuals in the study, the strength of the toxic substance, the amount of exposure to the substance, the number of other factors that might cause the harmful effect, and the length of time the people are studied. This information may take many years to obtain.

Business interests also want to "go slow" because correction of hazards costs money. Businessmen often argue that the cost of the required safety features will force them to close their plants. They deny that there are involuntary risks. They point out that many people are willing to undertake risks if they are paid enough. Management argues that all it has to do is explain the risks and then people can decide either to voluntarily take the risks or avoid them.

Businesses support risk benefit analysis that claims to provide objective ways to measure expected hazards against expected benefits of technology. They argue that regulatory agencies could then make politically neutral decisions and establish acceptable standards based on experience.

There is a fallacy in this argument. Those who are bearing the risks are in a weaker power situation than those imposing the risks. If women are in urgent need of money, job loss or reduction in take-home pay may be too high a price to pay for a reduction in health hazards. Women workers may also be reluctant to talk about health problems, particularly if they concern pregnancy, for fear that they may lose their jobs if they speak out. Based on previous experience, they have little faith in management's desire or efforts to minimize reproductive health hazards. Some might be willing to accept hazard pay to improve the financial security of their families under certain circumstances, but this trade-off only makes the choice between work and health even more difficult as it highlights the risks. If you are pregnant, this is really not an option.

Governmental regulatory agencies are caught in the middle. They are pressured on the one hand by the business and scientific communities who, for different reasons, advocate a cautious approach, and on the other hand by workers and consumers who demand that they act immediately. Each side uses research studies to buttress its own position. Perhaps we need to find a new way of looking at evidence of health hazards during the interim period when more conclusive proof is not yet available. The scientific standard of making no recommendation unless there is at least 95% certainty that a given substance is harmful may be too stringent a criterion to use when possible severe health damage is at issue. What makes sense for scientific purposes may not make sense when you are trying to protect your well-being and the well-being of your unborn child. If at some future date, scientists determine according to their acceptable statistical criteria that your workplace exposure could cause reproductive harm, these findings will be too late to help you.

WHAT RISK SHOULD YOU TAKE?

Society assigns mothers primary responsibility for the health of their fetuses. Thus, concern about the outcome of their pregnancies takes a high toll in anxiety as well as illness. In order to alleviate their fears, they need to learn how to rank hazards they face in terms of importance and decide which ones they are going to try to do something about and in what order. Without some basis for judgment, hazards they choose to avoid may not be those that do the most harm.

> When I was pregnant, I was concerned about the duplicating fluid. I would have purple fingers from the ditto and I was always concerned that it would absorb into my skin and get to the baby, so I always washed my hands carefully.
>
> –Kitty, teacher in a private school

Evaluating the risks we face may seem like an impossible task because of the amount of uncertainty involved, but research on the accuracy of public perceptions of health risks shows that consumers have a surprisingly good idea of the relative frequency of most causes of death, even though their knowledge about the risks may not be precise. They do tend to overestimate the hazards that receive a great deal of media coverage and to underestimate their own vulnerability. Pregnant workers are even more concerned about health risks than the general public and are likely to improve their risk estimating capabilities if presented with clear and accurate information.

What you decide to do with information depends on your opportunities. Experts cannot always tell you what the right choice is when you only have options that involve some degree of hardship or risk. They can only provide information. You are the one who has to live with your decisions. If you know, however, about the risks attached to reproductive hazards, you are able to make some kind of choice, even if the choice ultimately boils down to whether you stay on the job or leave. What is most upsetting though, is that hazards are often hidden. It is not until someone suffers the ill effects, that you learn that you too are at risk.

> I know of one incident where a woman was working with a lot of chemicals in the electronics company. She had quite a few miscarriages. Her doctor had to tell her it was related to all the chemicals she was working with. Two other pregnant women were not aware of any hazard until their doctor told them and then they did go out on disability. They had trouble collecting from the company which wasn't so willing to admit that it was more or less their negligence.
>
> —Caroline, electronic microcircuit wirer

Pregnant workers in these circumstances may decide to quit rather than face the risk of harming their unborn children. This is only a possibility for those lucky few who can live without their salaries for the nine months of their pregnancies.

> When I took X rays I was afraid of the scatter radiation. I wore a badge and every month we took it off and sent it out to the lab to make sure I wasn't getting any radiation. I was concerned about that. We tried to work it out that the other girl would take more of the X rays. Sometimes it worked and sometimes it didn't.
>
> The lab would send the information back on the badge whether it was negative or positive. It was always negative. If there was any trace of radiation, I would have had to terminate my employment.
>
> —Rory, dental assistant

Rather than quit, others attempt to avoid those aspects of their jobs they feel are dangerous during their pregnancies. If their supervisors are sympathetic and co-workers willing, this solution often works.

> I worked in the sterilizing department of a medical center and ethylene oxide was used for sterilizing. It is a poison gas. After I realized I was pregnant, I wouldn't work with the gas sterilizer. There was no problem with this from my supervisor.
>
> —Pam, works in the sterilizing department of a hospital

Still others ask for a leave of absence or a transfer to other kinds of jobs for the duration of their pregnancies. Next to cleaning up the workplace for all – the major goal – these are often the best options, but the least frequently available. Even when choices are open, it is often not clear which ones to take. This is not a very satisfactory state of affairs, but one we have to live with for the immediate future.

You must become a chronic question asker. The questions listed earlier in this chapter and on the following pages help you ask the right ones. The previous questions related to understanding articles written about reproductive health hazards. The next sets of questions pertain to both general reproductive health hazards in the workplace and those specifically geared to your own possible exposure. Despite the gaps in available information, these questions can help you decide how risk-averse you want to be.

GENERAL QUESTIONS TO ASK WHEN YOU ARE JUDGING REPRODUCTIVE HEALTH HAZARDS AND THEIR RISKS

1. Determine whether a substance is harmful to your health
   a. Can it damage the fetus?
   b. Can it cause a mutation?
   c. Can it damage specific organs of the body?
   d. Can it cause cancer?
2. If it is harmful, how harmful is it?
3. If it is harmful, what is the likelihood that you are exposed to it?
4. Who may be particularly at risk?
   a. Pregnant women
   b. Fetuses
   c. Individuals working in certain occupations
   d. Individuals living in certain communities
   e. Children
   f. Men and women contemplating having a child
5. What kind of evidence was the risk assessment based on and how strong is it?
6. What were the sources of information on which the estimates were based?
7. How much uncertainty is involved in the estimations?
8. Upon what assumptions are the estimates based?

FINDING OUT WHETHER YOU MAY BE AT RISK FROM AN OCCUPATIONAL HAZARD

Information Needed:

1. Find out what substances you are exposed to in the workplace.
2. What kind of health problems can be caused by these substances? Some individuals have allergies to substances that do not seem to have ill effects on others. If you suffer from any allergies and are pregnant, be sure to question carefully about possible kinds of allergic reactions.
3. Do you breathe (inhale), swallow (ingest) or have skin contact with these substances?

4. For how long have you been exposed to these substances – minutes, hours, days, months, years? Is the exposure continuous or occasional (sporadic)?
5. How high a dosage of the substance are you exposed to?

Sources of information:

1. Wear a dosimetry badge on your lapel that will measure air pollutants or obtain permission from your union or employer to periodically test the air in your workplace with a special kit that is made for that purpose. Then bring the sample to an independent laboratory for analysis.
2. Use the revised Federal Hazard Communications Standard (Right-to-Know Law). Many states have their own Right-to-Know laws. These laws give you the right to be informed about the chemicals in your workplace and possible toxic effects.
3. Ask your family doctor, employer, and union.
4. Ask your county or state health department, community medicine department at the nearest medical school, or public interest health group.
5. Write or call the Cincinnati office of the National Institute for Occupational Safety and Health (NIOSH). Address can be found at the back of the book.

SUGGESTED READINGS

British Medical Association, 1988, *Living with Risk*, Wiley, NY.

The Economist, 1985, "One Man's Acceptable Risk is Another Man's Accident," January 19, pp. 81-82.

Graebner, W., 1984, "Doing the World's Unhealthy Work," *Hastings Center Report*, 14:4.

Nelkin, D., and Brown, M., 1984, *Workers at Risk: Voices From the Workplace*, University of Chicago Press, Chicago.

Pagen, B., 1982, "Controversy at Love Canal," *Hastings Center Report*, 2:3.

Schneider, K., 1985, "The Data Gap: What We Don't Know About Chemicals," *Amicus Journal*, 6:3, Natural Resources Defense Council, NY.

TABLE 6.1. Some Substances Thought to Cause Adverse Reproductive Health Effects in Animals or Humans Due to Occupational Exposure

Exposure	Female	Male
LEAD		
Prior to conception	- possible chromosome aberration (h) - menstrual disorders (h)	- possible chromosome aberration (h) - sperm abnormality (h) - degeneration of testes (h) - decreased sex drive (h)
At conception		- sperm abnormality (h)
During pregnancy	- miscarriages (h) - stillbirths (h) - malformations (a)	
On newborn	- lead in breast milk (h)	
On child (through clothing)	- lead poisoning (h) - hyperactive (h) - brain damage (h)	- lead poisoning (h) - hyperactive (h) - brain damage (h)
ANESTHETIC GASSES		
During pregnancy	- miscarriages and birth defects (h) (female exposure and male partner exposure)	
ESTROGENS		
Prior to conception	- effect on flow, frequency of menstrual cycle (h)	- sore and enlarged breasts (h) - impotence (h)
On child (through clothing)	- sore and enlarged breasts of prepubescent (h)	- sore and enlarged breasts of prepubescent (h)
IONIZING RADIATION		
Prior to conception	- mutations in genetic material (a) - reduced fertility (h)	- mutations in genetic material (a) - reduced fertility (h)

TABLE 6.1. (continued)

During pregnancy	- leukemia and other cancers in offspring (h) - cataracts and eye defects in offspring (a) - small heads and brains in offspring (a) - mental retardation and retarded growth in offspring (h)	
	PESTICIDES	
At conception	- prevent conception (a)	
During pregnancy	- stillbirths (a) - miscarriages (a) - abnormal offspring (a) - leukemia in pregnant mother and offspring (a)	
On newborn	- pesticides in breast milk (h)	
	VINYL AND POLYVINYL CHLORIDES	
Prior to conception	- chromosome aberrations (h)	- chromosome aberrations (h) - mutations in genetic material in sperm (a)
During pregnancy	- cancer in offspring	- miscarriages and stillbirths in partners (h)
	BENZENE, TOLUENE, XYLENE	
Prior to conception	- damaged chromosomes (h) - menstrual disorders (h)	- damaged chromosomes (h)
During pregnancy	- aplastic anemia in woman (h) - birth defects (h)	
	CHLORINATED HYDROCARBONS	
Prior to conception	- chromosome aberrations (h)	- chromosome aberrations (h) - infertility (a)

During pregnancy	- liver damage to fetus (a) - in breast milk (h) - miscarriages (h) - cancer in offspring (a) - spinal birth defects (h) - stillbirths (a)
	CARBON MONOXIDE
During pregnancy	- smaller size of newborn and higher chance of dying soon after birth (a) - stillbirths, cerebral palsy, mental retardation in offspring (h)
	CARBON DISULFIDE
Prior to conception	- irregular menstruation (h) - extreme bleeding (h) - decreased sex drive (h) - sperm abnormalities (h) - impotence (h)
At conception	- decreased fertility (h)
During pregnancy	- miscarriages (h)
	PCB's
At conception	- reduced ability to become pregnant (a)
During pregnancy	- small babies with PCB's in tissue (a) - stillbirths (h) - babies born with discolored skin which then fades (h)
On newborn	- PCB's in breast milk (h)

KEY: (h) At least one study on humans showed this effect.
(a) At least one study on animals showed this effect.

Sources: Barlow, Susan M., and Frank M. Sullivan, 1982, Reproductive Hazards of Industrial Chemicals: An Evaluation of Animal and Human Data, Academic Press, London; Hricko, Andrea, and Melanie Brunt, 1976, Working for Your Life: A Woman's Guide to Job Health Hazards, Labor Occupational Health Program, University of California, Berkeley; U.S. Congress, Office of Technology Assessment, 1985, Reproductive Health Hazards in the Workplace, U.S. Government Printing Office, Washington.

Photo © Earl Dotter

Chapter 7

What We Have and What We Want: Self-Power, Women Power

We have come a long way since the "factory girls" labored under unsafe and unsanitary conditions for four cents an hour. As we have seen, though, subtle types of harassment and discrimination have accompanied these improvements. These hidden agendas are more difficult to prove and in some ways more difficult to combat, as the practices are defended on other grounds—efficiency, protection, equity—all of which sound laudable. Pregnancy or just the biological potential for pregnancy has been used in the past and continues to be used by employers as the unstated excuse for policies that impede equality, job security, and career mobility for women.

> The new principal was a real humdinger. She's a drill sergeant type of person and she particularly is interested in having the place clean and neat. At 8 months pregnant, I was on top of the chair dusting off the top of a coat closet and I thought to myself, "this is stupid." Then I thought that all the emotional stress that she put on me was bad for my baby. I feel she caused my high blood pressure problem. This woman obviously resented me being pregnant and on her staff.
>
> —Dinah, parochial school teacher

WHAT CAN WE DO?

Pregnant workers can improve their lot in many different ways. Some involve major changes in policy and expenditure and others only require small substitutions that employers might be quite willing to make if these were brought to their attention. Furthermore, pregnancy can become a catalyst for action. Women will often organize around reproductive threats in the workplace and community more readily than they will for other serious health hazards.

Throughout the country, women by themselves and in groups have exerted pressure on employers and the government to reduce health hazards on the job and in the community. Their efforts have received a good deal of local media attention, but with the exception of Love Canal, very little national coverage. Yet these efforts have made a difference. More women would organize for change if these stories were told and "know-how" transmitted. Women have been very ingenious and creative in the ways they have gone about improving the comfort and safety of their workplaces. Furthermore, they have developed techniques and approaches that can be used by the shy and retiring as well as the more assertive female worker.

In one office, women were having breathing problems which they suspected were caused by polluted air in their offices. They contacted the staff of the occupational safety and health program of their union, the Communication Workers of America (CWA). The CWA representative and the women involved investigated the problem together. First, they requested information from the employer's files about the chemicals used in the ventilation system. They discovered that freon, a highly toxic solvent, was being used. Then they conducted a simple survey. This documented the substantial number of workers affected and pinpointed the location of the workstations and air vents.

Next, the CWA brought in a health expert to conduct tests. Backed by the results of the tests and their own evidence, the women conducted informational picketing in front of the building. This led to bad publicity for the company. Management soon relented and installed a new, more efficient ventilating system. Most importantly, the new system used a less toxic solvent.

In another office, the CWA, by filing grievances with OSHA, convinced the employer to finance a health and safety education program. The women wanted to learn how to identify hazards themselves so that these could be corrected immediately. By eliminating the time lag involved when the governmental agency was called in to investigate and confirm the hazard, the women would hasten the process. The company also agreed to this request because it preferred not being the target of government inspectors.

After taking the health and safety classes, one group of women was able to identify ozone emissions from a copying machine. They tried to persuade management to switch to another type of copying machine designed to prevent ozone formation. They were not successful, but did get management to improve the ventilating system, a second best solution.

One of the most effective demonstrations of women power in the office occurred at Boston University's Business School, where members of Local 65 of the United Automobile Workers refused to unpack the new video terminals until management provided ergonomically sound workstations similar to those installed in other departments. The unpacked boxes stood there for six months until the women's needs were met.

Sometimes pregnant workers are able to accomplish a good deal on their own. The National Action Committee on the Status of Women has collected several case histories. For example, Tess, a pregnant laboratory technician, worked in a non-unionized battery factory. Her job was to test samples of soil from the work site. When she discovered that they showed mercury contamination, she asked her doctor about the risk to the fetus posed by her exposure to mercury vapor. Her doctor warned her not to have any more exposure to mercury throughout the rest of her pregnancy because mercury vapor can be transported across the placenta inflicting possible damage to parts of the developing brain and nervous system of the fetus.

Tess requested a transfer to a mercury-free area of the plant. She was offered several positions that had less mercury exposure, but none at a zero level. She refused these. The company responded by asking her to sign an indemnity form which absolved the company from any responsibility for harm to herself or her unborn child if she continued working. Tess refused to sign. She finally negotiated a transfer for the remainder of her pregnancy to an office job which had zero exposure to mercury.

Tess's story had a happy ending, but sometimes we ask the right questions but get the wrong answers. Take the case of Diana, whose Canadian employer embarked on extensive renovations involving paint varnishes and glues in the area in which she worked. When she asked if these could be harmful to her pregnancy, she was assured that there was no danger. Three weeks after the work was started, Diana miscarried. She requested a copy of the data sheets listing chemical ingredients to show her doctor. Her physician felt that the presence of toluene in the paint and exposure to several other chemicals had caused her miscarriage.

Diana felt particularly betrayed because she had repeatedly asked for information about the fumes and was repeatedly assured that they were harmless. She then charged her employer with violation of the *Canadian Occupational Health and Safety Act* (the Canadian Act allows individual prosecutions). These legal remedies will not bring Diana's unborn child back, but they should make the employer more careful in the future and help protect employees' future pregnancies.

Cases like Diana's show why a Right-to-Know law is so important. But the law is only valuable if it is used. The laws, unfortunately, are often hard to understand and difficult to use. This is one area where women can put pressure on the government to improve the clarity and usability of the legislation.

When work situations are appalling and discrimination rampant, a courageous woman can find the strength to fight back under incredibly difficult circumstances and become a pathbreaker for the rest of us. Bonita is one such heroine. As a pump tender at the Steel Company of Canada, she suffered six years of unbelievably vicious harassment simply because she was a woman in a traditionally male job, and because she often filed complaints about safety and health hazards. For example, a gas booster that she complained was hazardous later exploded. Instead of being applauded for her actions, she was called stupid, accused of theft, her work competence was questioned, and she was threatened with layoffs and the elimination of her job. She was used as a scapegoat to intimidate the remaining women at the plant into quitting. The company sank so low in its tactics that it refused to screen off the men's shower and changing room. Bonita was forced to see the men naked day after day. When she complained, she was told by her supervisor to shut up and enjoy the free show!

Bonita did not enjoy the free show. Instead she went public and filed charges against the company, listing over 60 instances of abusive treatment, discrimination, sexual harassment, reprisals from management, and unhealthy and unsafe working conditions. She demanded a far-ranging series of remedies and her complaint is precedent-setting. It seeks both to get the health and safety legislation to cover sexual and gender harassment and to compel companies to properly accommodate their facilities and procedures to women workers in nontraditional job areas.

Not many of us could persevere under the conditions Bonita faced. Luckily, they represent only one end of the spectrum—an end that shouldn't exist. But most women are capable of proposing some small innovation to improve their working lives during pregnancy. They can then build on their increased knowledge, competence and self-confidence. They can start in a small way in an area that is not likely to make waves and one in which their non-pregnant colleagues concur.

For example, after decades of unconcern, Americans have adopted health diets and physical fitness in a big way. Many large corporations are offering recreational programs to improve the health of their workers and reduce absenteeism. You can benefit from this trend.

Here are a few suggestions. If the workers in your company are requesting such a program, why not ask them to include an exercise class geared to pregnant women? Why not also ask for the cafeteria to be open for breakfast a half hour before the start of the workday, and for it to provide fresh fruit, salads, juices, and milk if it does not do so already.

Often you are in such a hurry in the morning, you do not have time to eat breakfast at home. Instead you snack on coffee and cake at break time. If your employer does not want to get involved in the food business, suggest he look into the competitive food service companies which provide meals and snack food on a contractual basis. At the very least, your employer can provide you with vending machines that carry sandwiches, fruit and yogurt.

Many of you work in small offices and businesses that cannot afford to hire food services or open a cafeteria. At a minimum expense these businesses can furnish a small area with comfortable chairs, good lighting, hot and cold running water and a small refrigerator, which can double as a lunch room and employees lounge. You can then put up your feet, reduce the pressure on your legs, and relax. Socializing in pleasant surroundings reduces tension and fatigue.

You can also request your union to include flexible break schedules for pregnant workers in contract negotiations. If you have no union or formal contract agreement, you can suggest to your employer that you take shorter lunch breaks or come to work 15 minutes earlier in exchange for an extra break during the day. Another way of handling this is to use personal or sick day leave for this purpose. One eight-hour personal day would provide 32 days with an additional 15-minute rest period, a valuable exchange.

Pregnant workers need a mandate setting down the desired goals. Then tactics toward reaching those goals can be devised.

Pregnant Workers' Mandate for Employers

1. Top managerial commitment to eliminating or reducing reproductive hazards in the workplace
2. Incorporate work and family-life commitment into company policy and public statements
3. Establish hazard monitoring procedures and a grievance process
4. Investigate the feasibility of instituting any or all of the following policies:
    a. Temporary transfer of pregnant worker or male and female worker trying to conceive from jobs having possible reproductive health hazards, maintaining job security and equivalent pay
    b. Support research efforts to assess reproductive hazards of the workplace
    c. Clean up the workplace if known hazards exist
    d. Train the workers regarding health and safety procedures
    e. Provide: protective clothing that fits, sufficient ventilation, adequate heat, accessible and comfortable lounge and eating facilities
    f. Provide flexitime schedules
    g. Provide on-site child care center
    h. Provide child care referral services

i. Provide part-time work
j. Provide job sharing
k. Provide parental leave and benefits
l. Provide special personal days or time off for pregnancy check-ups and family obligations

Pregnant women can fight for occupational health, pregnancy benefits, and child care facilities on several levels: as individuals, with a small group of co-workers, as union members, as part of a community health group or political lobbying effort. These initiatives require a great deal of time and energy. Pregnant working women and working mothers of small children have little extra time or energy. Therefore, informal alliances between women's health groups, unions, COSH Groups (Coalitions for Occupational Safety and Health), and consumer environmental groups have been formed to press corporations and the government regulatory agencies to work toward the commitment to a healthy reproductive workplace and a healthy work and family policy for all.

Before you try any of the initiatives suggested in this chapter, it is valuable to have a model of what an ideal cooperative arrangement between employers and employees would be able to accomplish in protecting the reproductive health of both male and female workers. Sweden is one nation that is way ahead of the United States and the Swedish Chemical Industry has introduced such a prototype. Progress on workplace protection of reproductive health throughout the world has been slow. Considering, however, that people have been aware of reproductive hazards of the workplace at least since Roman days, but nobody has been concerned enough to do anything about them, present day policies in a few industrial countries signal a change in priorities.

A MODEL FOR THE UNITED STATES

The Swedish initiative started in 1977 when a female safety delegate in a quality control unit pointed to a suspected increase in the number of spontaneous abortions and birth defects among women employed in an industrial laboratory. A study of the problem was undertaken and an outside expert was immediately called in to participate in the inquiry. This study looked at the past histories of miscarriages and malformations among women working in laboratories and among unexposed women working in offices of pharmaceutical companies.

A preliminary report issued in 1980 supported the plausibility of the thesis that spontaneous abortions and congenital abnormalities were somehow linked to the women's workplace exposure to chemicals. But as we saw in Chapter 6, the number of women in these epidemiological studies is often too small to be able to determine with statistical significance whether the substance being investigated is responsible for the harm or whether the findings could be explained by chance alone. More importantly, the study lacked important information about the levels of chemicals the women were exposed to and under what conditions.

Instead of what would be the likely response of American industry—to study the problem further before instituting remedial action—the Swedish chemical industry initiated plans to substantially improve the working environment in chemical laboratories *even before the results of the report were ready*. They also introduced a required safety course for all employees on chemical risks in the lab and permitted susceptible workers to transfer to work that did not expose them to these risks.

In addition, the chemical companies' health services were directed toward preventing new work-related reproduction disturbances. In some instances, an entire chemical process was viewed as being a chemical risk environment, whereas in other cases only specific phases of work were considered to be hazardous. A combination of active cooperation from management and the appointment of women as special contact persons has aided in making the program work.

In 1984, partly in response to the earlier study, the Association of Swedish Chemical Industries published a set of recommendations for the industrial handling of chemical substances that possibly affect human reproduction. The suggestions were aimed at all segments of the industry: to the industry as a whole—to the individual company, line management, personnel sections, and company health services—and to local trade unions. American companies need to be governed by similar standards which could be adapted to fit non-chemical reproductive hazards in the workplace as well.

Some of the main points of the recommendations are:

To the chemical industry:

1. To strive for the best possible working environment in all workplaces.
2. To make no *general* separate regulations for men and women. For most substances, no one can state with confidence that the risk is greater for women than men, or vice versa.
3. To maintain a good interchange of information between research and practice. This is necessary so that scientific findings can be speedily implemented on the job and research can be initiated when practical knowledge indicates the possibility of a hazardous condition.
4. To develop relevant testing and screening methods.

To the individual company:

1. To survey each workplace regarding exposure to substances potentially damaging to reproduction.
2. To minimize the risk for harmful exposure to these substances.
3. To obtain and properly assess available information on potentially harmful reproductive effects of chemicals being handled.
4. To prevent or limit exposure of particularly vulnerable groups where required.

Line management is responsible for overall policies regarding:

Risk inventories and handling instructions in the workplace; protective measures; information, particularly for new employees, and necessary transfers to other jobs.

Personnel tasks:

Planning health information programs, especially for new employees, and formulating procedures for job transfers when necessary; participating in preventive measures such as providing appropriate information (which will not lead to loss of job) to the health service of the company; considering the legal rights of the workers when making decisions.

Local Trade Unions:

Providing information to personnel, particularly with reference to new employees; working out election procedures for special union representatives for those work sites where reproductive hazards are suspected.

Company Health Service:

Undertake workplace surveys, risk assessments, medical examinations, and reproductive history of new employees; provide health examinations which systematically follow up reproductive disturbances; when necessary inform workers as to what cautionary measures to take; measure atmospheric concentrations of substances in areas where reproductive disturbances are suspected; furnish pregnancy information, and, if possible, systematic follow-ups of pregnancy outcomes in the worksites where workers may be at reproductive risk; issue regular reports to the safety committee.

(Olin, R. et al., 1986. Industry in Sweden: a program of action. *Occupational Medicine: State of the Art Reviews* 1;3:483-495).

PATHS TO PROGRESS

The Swedes have established the kinds of reproductive health standards which we all want and which we hope we will eventually receive. But, unfortunately, the U.S. is still far from reaching such a goal. Along with the striking gaps in knowledge about reproductive health hazards in the workplace are the gaps in communication and cooperation created by the parties that should be working together to alleviate the dangers—workers, affected communities, labor unions, employers, clinicians, researchers, and policy makers.

There are, however, many paths toward progress. One or more of these routes may fit your particular workplace situation. The greater the success on local levels throughout the country, the sooner the country will be ready to emulate the Swedish model.

Organize Around the Issue Involved

A small group of workers can effectively put direct pressure on their employer by finding commonalities with other workers so pregnancy is not seen as an aberrant condition. Men want their wives' pregnancies to be protected and non-pregnant women want their future pregnancies or those of their family members to be protected. Remember that if a workplace substance harms reproduction it also has other harmful effects on the health of all workers. Management is also not immune to these risks if they too are exposed on the job.

Get Your Union Interested and Have Them Negotiate Workplace Improvements into the Contract

For example, Local 65 of the UAW at Boston University has a health and safety article in its contract that reflects the concerns of the pregnant VDT workers:

> Although research to date has not proven that video display terminals are a health or safety hazard, in recognition of employee concern about the potential adverse effects involving pregnancy, the University agrees to the following: Upon request, the University will attempt to reassign a pregnant employee to work which does not require the use of a video display terminal. If such reassignment is not practicable, the employee, upon request, may be granted a Personal Leave up to three months under Article 39, which shall be in addition to any other leave under Article 39 for which she may be eligible.
>
> Article 18 of District 65 UAW Contract at Boston University

Put Pressure on Regulatory Agencies to Change Their Standards or Enforcement Policies

Ideally, the workplace should be cleaned up for everyone so that men as well as women do not suffer reproductive harm. The government has passed laws and set up regulatory agencies (see Appendix A) to ensure healthy workplaces and environments. Unfortunately, these agencies are often underfunded, understaffed, and undermined. They have great potential for improving conditions, however, if the public demands such action and elects officials sympathetic to this goal. Federal, state, and local Communication Hazard Standards, better known as the Right-to-Know Laws, are recent additions to the pregnant worker's arsenal for self-protection.

Regulatory agencies should be willing to consider that an exposure is dangerous when only a small number of studies indicate harm. This could be done by passing new regulations under which OSHA can issue emergency temporary standards on such a basis. Amendments to the OSHA act are much needed. One overdue amendment would allow employees to act against employers who are thought to be violating health and safety regulations. Another would make OSHA legally bound to respond to NIOSH recommendations concerning reproductive and other occupational health hazards. A third would be to give the workers the right of private action, which is the right to bring OSHA to court for alleged violations.

This is the time to press for changes. The original OSHA legislation has been in effect for over twenty years. For a long time, interested parties were opposed to opening the Act for fear of unwanted amendments. Times have changed and several bills aimed at reforming OSHA have been introduced in Congress. Some limited reforms are now possible.

Find out what protection you already have. Determine whether it would be more effective to ensure that the regulatory agencies enforce existing policies or to lobby for new regulations. One way the Reagan administration has emasculated enlightened environmental and occupational health policies has been by withholding money for enforcement or appointing administrators with anti-regulating views to implement enforcement procedures.

While enforcement is to a great extent in a state of suspended animation, OSHA continues to publish pamphlets about worker health and safety and has recently revised its pamphlet on *Employee Workplace*. You may also find useful *Your Workplace Rights in Action* and *All about OSHA*. They can be obtained from the OSHA Publication Office, Room N-3101, 200 Constitution Ave., NW,

Washington, DC 20210. Owners of small shops and offices usually do not have sufficient information about worker rights, so these booklets are especially helpful if you work in one of these small companies. Many government agencies offer toll-free information lines. Check with the toll-free information operator in your area (1-800-555-1212). The EPA runs several of these services. You can receive a variety of nontechnical information material from the EPA Public Information Center. More technical information can be obtained by calling (202) 829-3535 (not a toll-free number) or by writing to PIC (PM-211B) U.S. EPA, 401 M Street, Washington, DC 20460.

Have Your State Pass New Laws or Have Old Ones Amended

Lobby for a law requiring employers to allow workers attempting to conceive a child or who are pregnant to leave hazardous jobs temporarily without loss of pay. Another helpful law would require employers to provide scientific evidence that no harmful effects are transmitted from the exposed father to the offspring before they can establish a fetal protection policy in their plants. Legislation prohibiting discrimination on the basis of genetic characteristics, reproductive capacity, or other physical or biological conditions that are not connected to the ability of the person to perform the job is also badly needed.

Urge your state legislators to amend the workers' compensation system so that reproductive health is covered. Right now only the worker is covered. Since the fetus is not the employee, Workers' Compensation laws do not consider damage to the unborn to be under their jurisdiction. Second best would be to provide the *right* for workers to pursue a tort remedy for reproductive damage not covered.

Get Involved in "Barefoot Epidemiology" Where Women Affected by Toxics Gather the Evidence of Harm Themselves

The most well-known cases of barefoot epidemiology were undertaken at Love Canal where the women refused to accept New York State's conclusions that living in their homes did not pose any reproductive risks, and in Woburn, Massachusetts where women traced an increase in children's cancer and leukemia to toxic wastes from two major industrial companies.

A more recent example involved a woman pharmacist in a small town in Louisiana who became concerned because her sisters, friends, and neighbors had become pregnant at the same time and all had miscarried. When she finished her inquiry, she found that one out of three pregnancies in her community did not end up in a live birth, more than twice the average for Louisiana. Her list was taken seriously because the town she lives in lies in the midst of an industrial corridor between Baton Rouge and New Orleans producing one-fifth of America's petrochemicals and having one of the highest cancer rates in the country. The air, ground, and water are so full of carcinogens, mutagens, and embryotoxins that a union leader pithily referred to it as "the national sacrifice zone."

A few American, Canadian, and British studies show an abnormally high rate of spontaneous abortions and birth defects among women whose husbands' work exposes them to vinyl chloride or who live downwind from vinyl chloride polymerization plants, which was the situation in Louisiana. Scientists at the Tulane School of Environmental Health are now searching for the reason behind this high incidence of miscarriages.

Men and women in workplaces exposed to potentially dangerous substances have also generated important hypotheses about reproductive hazards through their own experiences. Workers' self-reports may be the first indication that anything is wrong. In fact one union leader asserted that in all the cases he knew about, experts called in to investigate upheld the workers' contentions that there was a problem.

Probably the most dramatic case of worker-reported reproductive difficulty occurred in the late 1970s when several seemingly innocuous conversations led to the eventual banning of the chemical DBCP because it was toxic to the testicles. Workers in a California agricultural chemical plant and their wives noticed that no one in the group was having babies. It soon became evident that this was not by choice but a result of the inability of the couples to conceive. The workers became worried and informed their union. The union contacted university researchers requesting that they investigate the situation. The researchers confirmed the association between exposure to DBCP (dibromochloropropane) and decreased sperm count. The testicular toxicity of DBCP in animals had been known for more than 15 years, but without the couples themselves pushing for an investigation, it is unlikely that studies demonstrating the human connection would have been initiated.

Another group of workers exposed to halogenated hydrocarbons questioned what they thought to be an unusually large number of infant deaths. A preliminary epidemiological survey tentatively confirmed this connection.

Press for Research Funds to Provide the Information Needed About Reproductive Risks in the Workplace and Their Amelioration

The setting of research priorities, the allocation of money to such investigations, and the design of acceptable studies are often impeded by political, scientific, and economic obstacles. Researchers of women's health problems, particularly from a women's perspective, have an unusually hard time. The funding agencies, despite evidence to the contrary, are not convinced that women will maintain the regime required over a long period of time, whether it involves eating a low-fat diet in a breast cancer study or taking monthly pregnancy tests in a VDT and reproductive health investigation. Most of the funding sources are dominated by males who do not seem to realize that, for women, diligence becomes an especially high priority when such traumatic and as yet intractable problems are involved.

Clusters of workplace-related harm always need to be investigated carefully. In the past, analyses of clusters of abnormalities confirmed their relation to pharmacological, workplace, and environmental hazards. The teratogenic properties of methyl mercury, rubella, thalidomide, and DES (diethylstilbestrol) were documented by this route.

Clusters of miscarriages and birth defects among VDT operators have been the impetus of VDT use and reproductive health studies in several countries. Though there is no solid evidence linking VDT use to reproductive harm, two studies have shown a link between miscarriages and working more than 20 hours a week at a video terminal. Whether these results will hold up in further studies, or whether some aspect of the machine, stress, or some characteristic of the workers is responsible for the findings is not yet known.

Go to Court

Sometimes going to court is the only path open to obtain redress. In 1990 in Tampa, Texas, a lawyer initiated the country's first legal test of a medical theory connecting the exposure of parents to toxic chemicals with the birth of a child with Down's syndrome (a major cause of mental retardation). The case was brought against a Hoechst Celanese chemical plant which was accused of contaminating the air and drinking water of the community. Some medical researchers believe that toxic chemicals could damage proteins and enzymes involved in the separation of chromosomes at the point when either the egg or sperm are formed. If this is the case, it is possible that an extra chromosome would result. Down's syndrome is usually caused by an extra chromosome 21. While the first legal battles may not be won, increased evidence linking male exposure to toxics with reproductive damage should eventually translate into legal victories.

Even if the court option is the only avenue of redress, it is not a good choice for many women. It is a stressful, expensive route with an uncertain outcome. In fact, your employer may be counting on these factors to dissuade you. But if there is no alternative and you are adamant about pursuing justice, then the court may provide a viable choice. Take the case of a female sales representative who had suffered two miscarriages. After the second one, she was told that she was being reassigned to an inferior territory to make room for a male representative who was being transferred to her job. When she complained to senior management, she was given an ultimatum: accept the less lucrative assignment or resign. She did neither. Instead she went to court and won her case. The judges ruled that she had been discharged because of her sex and her expressed desire to combine motherhood with her career (*Goss vs. Exxon Systems*).

Plaintiffs can also file pregnancy discrimination charges under

Title VII as amended by the Pregnancy Discrimination Act (PDA). Several thousand of these are filed each year, but according to the Equal Employment Opportunity Commission (EEOC), few involve allegations of discrimination on the basis of reproductive risks. This is probably because the EEOC does not have a good record on these types of cases.

Court cases can be brought by groups as well as individuals. Groups have more resources, both in terms of money and expertise, and are likely to become involved in class action cases. In the environmental field, two consumer groups, Public Citizen, (Ralph Nader's group) and the National Resource Defence Council (NRDC) have been very active.

The NRDC brought a case involving vinyl chloride emission in the production of plastics. In July 1987, the U.S. Circuit Court of Appeals ruled 11-0 that the Environmental Protection Agency (EPA) can only use health factors to determine safe emissions levels of toxic pollutants. This was a breakthrough, as previously the cost to the company of the clean-up was included in the determination of a safe standard. Vinyl chloride is thought to be a human carcinogen and to be toxic to embryos. It is high on the list of suspected chemicals contributing to the "epidemic" of miscarriages in Louisiana's petrochemical corridor.

Join a Non-Profit Volunteer Community Action Group or Form Your Own

Some community groups seek legal remedies, others favor the legislative route and focus on political lobbying. But the majority emphasize education, organization, and working directly toward a specific goal. One of the most successful, the Citizen's Clearinghouse for Hazardous Waste (CCHW), is built on "People Power," the community organization lesson learned from Love Canal. Basically their philosophy is to figure out who has the power to accede to their demands for a clean environment and to apply pressure until they achieve their goal. CCHW has passed this formula and its "know how" to over 1,000 grassroots groups. They claim that groups that were based on People Power won more than 90% of the time, whereas those groups depending on lawsuits and lobbying won less than half their cases.

Injured Workers United, Santa Clara Center for Occupational Safety and Health (SCCOSH) is one of many smaller local groups. The group is composed of a coalition of workers, unions, community activists, and health and legal professionals whose aim is to help impaired workers secure compensation and to fight for health and safety improvements. The term "injuries" includes harm to reproductive functioning.

(Several references dealing with organizing techniques are listed at the end of this chapter.)

Often the same pollutants that are affecting pregnant workers within the workplace are affecting the people who live near the worksite. An example of this occurred in the San Francisco Bay Area. Residents in a Silicon Valley community became concerned about an increase in spontaneous abortions and congenital birth defects. They organized and put pressure on the California Department of Health Services to trace the cause. The state investigators discovered that the local drinking water had been contaminated with solvents from an industrial storage tank leak.

If you suspect such contamination in your area, contact the environmental health division of your state health department. If you do not get satisfaction, contact The Toxics Coordinating Project (TCP), 2609 Capitol Ave., Sacramento, CA 95816, (916) 441-4077, which presses for investigations of toxic complaints and their cleanup, and keeps "toxics" as part of the political agenda. The TCP is part of the Coalition on Environmental and Occupational Health Hazards and has links with other groups around the country which may be able to offer direct help.

BARRIERS TO PROGRESS

Before pregnant workers can successfully win their case, they must determine what hurdles they have to overcome and decide how to get past them. One of the first steps for many women is to develop the confidence and skills needed to win over the opposition. Whether they choose the route of individual action or participation in group endeavors, some may need assertiveness training before they achieve this goal.

Many women believe strongly in gender equality. But when they start to take a forceful stand, they begin to feel uncomfortable. They realize their deeply held views about equality cover up images of sex roles that they learned as young children. They are not aware that these early images still exist deep inside themselves until they begin to behave in assertive ways. Will people feel I'm unfeminine or insensitive, or overbearing, ruthless, and hostile? All these stereotypes come to the forefront and the fledgling asserters begin to feel guilty. Nonsense! None of these characteristics has anything to do with assertiveness. Remember being feminine, loving, kind, understanding, and sensitive is not equated with giving up your rights.

A Woman in Your Own Right, by Ann Dickson, offers many good pointers and can function as a do-it-yourself guide. You can keep re-reading it when you slip back into old feelings of inadequacy and guilt. Here are some of the basic premises that this and other books on assertiveness training present.

What Does Being Assertive Mean?

- Deciding what you want and saying so specifically, directly, and calmly and sticking to your position.
- Preventing the person or persons you are confronting from intimidating you into feeling hostile, guilty, or inadequate. If they make you angry, express that feeling directly.

What Are Your Rights?

- The right to state your own needs and set your own priorities.
- The right to be treated with respect and dignity and to be considered an intelligent and competent person.
- The right to express your feelings, opinions, and values and to fight for what you feel you are entitled to.
- The right to make decisions for yourself and to make your own mistakes.
- The right not to be subjected to undue persuasion, bullying, or use of guilt to get you to change your course of action.

- The right to say you do not understand and require additional information and time to come to a decision.
- The right to say no or change your mind.

Role-playing is the most useful technique in learning to become comfortable about being assertive. Ask a good friend or co-worker to role-play with you and then give each other feedback. Simulate a real situation involving pregnant workers and management and you'll be surprised that after the first few awkward minutes, you will be feeling the part.

Start with an easy case first and then build up to the more difficult ones. For example, suppose that you want to bring your grievances about suspected reproductive hazards in the workplace to the attention of senior management. You think you will be able to get a more sympathetic hearing if you can get some supervisory personnel on your side. You know of one person that you might be able to convince. Start your role-playing with this person in mind. Take the part of the assertive advocate and have a co-worker who knows the supervisor in question take the supervisor's role.

In addition to being clear and specific about your demands, you must make sure that your body language does not cancel out your assertive verbal message. One common mistake in the early stages of assertiveness training is to concentrate so hard on what you are saying and how you say it that you forget to think about your tone of voice, facial expression, and posture. Once you are really *feeling* assertive, the assertive body language will follow naturally. But meanwhile you may have to unlearn some counterproductive body habits.

What Not to Do!

- Do not slouch, shift your weight from one foot to another, keep your head cocked to one side, fidget or shuffle your feet.
- Do not avoid eye contact or convey hostility or timidity.
- Do not paste a false smile on your face or grimace.
- Do not mumble, use a simpering little girl's voice, or use an apologetic or sarcastic tone.

What to Do!

An observer or your role partner can point out if you make any of the above mistakes. These can usually be corrected by readjusting your voice or body movement only slightly.

- Take several deep breaths to calm yourself before approaching a confrontation.
- Sit or stand in a well-balanced position, close enough to catch the attention of your audience.
- Speak slowly and distinctly. Modulate your voice varying the pitch from the high to low registers of your vocal cords.
- Dress in a way that makes you feel good about yourself!

Whether you read a book on assertiveness training or take a course, it is a good idea to role-play a "dry run" first to make you feel more secure and to strengthen your approach. You can even videotape your presentation and play it back.

There are a number of good references on assertiveness training and effectiveness:

A Woman in Your Own Right, Updated and Revised, Anne Dickson, Quartet Books, 1988.

Assert Yourself, Gael Lindenfeld, Thorsons, 1988.

How to Survive as a Working Mother, Lesley Garner, Penguin, 1982.

Self-Assertion for Women, Pamela Butler, Harper and Row, 1976.

Women's Rights: A Practical Guide, Anna Coote and Tess Gill, Penguin, 1977.

A few adult education programs offer a class in assertiveness training and unions sometimes include it as part of their education and organizing programs. Check with your union local and the educational institutions in your community.

In addition to lack of assertiveness and undervaluing your own ability, there are additional barriers that have to be overcome before successful organizing can take place. These are:

1. Lack of factual information
2. Not being able to pinpoint the problem
3. Not knowing who to go to in order to get the problem fixed
4. Belief that nothing will change anyway.
5. Overestimating the ability of scientists to know the answer
6. Fear of job harassment
7. Having promotion opportunities blocked, or being fired

This chapter and previous ones should provide you with both the information and additional resources that you need to overcome these barriers. You can then decide what efforts you wish to undertake in your own workplace and what regulations or legislation to seek from the state and federal government in order to reduce the reproductive risks in the workplace and compensate for prior harm.

The Office of Technology Assessment (OTA) of the United States Congress has issued a report, *Reproductive Hazards of the Workplace* (Washington, DC: U.S. Government Printing Office, OTA-BA-266, Dec. 1985). Women can build on the number of good suggestions OTA presented for future action that would protect the reproductive health of male and female workers and their unborn and live children as well as protecting employment rights.

ORGANIZING STEPS

Choosing the Issue

Gather information and identify what problem areas require correction. Survey the workers regarding their concerns and ascertain whether or not the suspected workplace reproductive hazard is considered serious enough for the women to be willing to act.

Be sure that you have some evidence to back your demands. This evidence can either be from a scientific study or from workers' reports regarding a suspiciously high number of reproductive health problems in your own workplace or others. It is essential that you have some feasible solutions in mind and that the women believe that they can do something about correcting the problem.

Choosing the Type of Action

Choose the type of action that the women feel comfortable in undertaking and maintaining.

Attempt to judge how hard your employers will fight back and what compromise they might settle for. Decide whether or not you are willing to settle for half a loaf.

Try to figure out what price you are likely to pay for your action if it is not successful. Let the women decide whether the chance of success is worth the penalty of failure.

Preparing for the Action

Make sure you have provided enough information and involved enough women in planning the action. Women need to believe that they can win and be reassured that they can effectively perform the disagreeable tasks that are part of the process of exerting pressure. Delegate work and motivate the fainthearted. Share the decision making. Individual strengths can be utilized and individual weaknesses counterbalanced in a group effort. People are more likely to feel stronger about an action if they have a part in designing the strategy.

Maximize the Threat Before Initiating Action

Be sure that your demands are clear and feasible and that you have ranked them in terms of importance. If in the bargaining process some of the demands have to be given up, there should be agreement on what these are beforehand.

Organize the Action Thoroughly

It is essential that each woman knows what she has to do and is willing and able to do it. If you cannot deliver on your threat, your future bargaining power is strongly diminished.

Plan to Publicize Your Victory in Order to Strengthen Your Position in the Future

Have press releases written. Appear on local TV and radio. Make sure personnel in other divisions of your company learn about your successful action. Inform other unions and safety and health committees, your local legislatures, state health departments, and any local community action groups and women's groups that might be your natural allies.

You may not want to become a lobbyist or an active organizer yourself, but being aware of the essential techniques enables you to make helpful recommendations to those who have undertaken these tasks and to warn against positions that are ill thought out or unworkable. An informed work force improves the likelihood of success. Suggestions for influencing the legislature and specific organizing tactics are listed at the end of this chapter.

THE FUTURE

Public concern voiced by women in the workplace and the community has raised society's awareness of occupational and environmental reproductive hazards. Awareness that the basic disagreements revolve around questions such as what is considered evidence and what policy decisions should be implemented is less widespread. Safety and profits too often clash, but this bottom line is covered with rhetoric largely cloaked in scientific terms. When millions of dollars are at stake, companies become defensive, callous, and insensitive. The pinnacle of insensitivity was achieved by representatives of the Louisiana chemical industry, who defended what they considered to be their exemplary environmental record. When asked to respond to the assertion that exposure to vinyl chloride from the petrochemical plants caused the high rate of miscarriage found in those living nearby, they answered that there is as little proof that chemicals cause spontaneous abortions as there is that "screwing" too much is the cause.

With such enemies, we need a lot of friends!

We Shall Overcome

Yes, we need a lot of friends! But we already have lots of friends—for the best friends we have are other women. Women in cooperation and coalition with each other and with concerned men will overcome resistance to achieving a pregnancy-friendly workplace. The road ahead is still a long and hard one, but it should be easier than it has been in the past. As we have seen, some progress has been made. The issue of reproductive hazards of the workplace has been raised nationally in the press and in broadcast many times. The Supreme Court decision on *UAW vs. Johnson Controls* made front pages of newspapers all over the country. The image of woman as sole vehicle of reproductive damage has been, if not demolished, at least questioned. Males themselves are beginning to realize that their reproductive functioning can also be damaged by workplace toxics and processes. Child care and family leave agendas are now part of social policy negotiations on the federal and state levels.

Women concerned about environmental and occupational health and reproductive hazards need to participate actively in all stages of the societal decision-making process on workplace hazards and risk. They are not "outsiders" to the process, but instead are "insiders" because they share the risk. More importantly, women are "insiders" who have been unethically disenfranchised from the decision-making process and are seeking redress. They must become extremely well informed and politically astute in order to attain a powerful role.

Hopefully in the future, women will be treated as equal partners in the risk evaluation dialogue and their viewpoints will be taken more seriously. Only then can women, together with scientists, regulatory officials, business leaders, health care personnel, and the public achieve a reproductive-friendly workplace for all.

CHOICE OF TACTICS FOR IMPROVING WORKPLACE

Your choice of tactics partly depends on the public relations and goodwill costs involved in fixing the hazard as well as the monetary cost of cleaning up the workplace. By clever use of publicity or the threat of going public, you can sometimes raise the cost of the non-monetary factors high enough for the employer to feel it is cheaper to expedite the safety measures.

Here are some approaches you can try that other groups of workers have used successfully.

Group Education

Hold these educational sessions at lunchtime so that workers from different parts of the company who are upset about the suspected reproductive hazards can come together and receive the same information. This reduces the misunderstandings that sometimes happen when different people provide information at different times.

Threaten Publicity

Prepare a flyer publicizing your grievances. Show management the flyer you are planning to distribute before you actually circulate it. Tell management that you will give them a chance to institute remedial action within a specific time frame before you implement your plan.

Use the Media

Local newspapers, TV, and radio stations are interested in occupational and environmental hazard stories, especially if you can make them sound dramatic. Work-related fertility problems, miscarriages, and birth defects find a ready audience.

Conduct Well-Directed Informational Picketing

Address it to the workers, to the public, to the company, and to politicians and civic leaders. Sometimes merely making information public and visible becomes a catalyst for action.

Innovative Actions

The magazine *American Labor* published an issue (No. 15) on "Do-It-Yourself Tactics: Local Action On Job Safety" which suggests several approaches that could be adapted for action on reproductive health issues. Some possible adaptations are:

Nurse-Out

Encourage all workers, male and female, who suspect that they are exposed to reproductive health hazards in the workplace to visit the plant nurse with complaints during working hours. Substances that are capable of inflicting reproductive damage usually have other harmful effects as well.

Lunch-Out

When the weather is warm, have lunch together on the company grounds at a at a site visible to the public. Bring large signs stating your grievances or fly helium balloons printed with an appropriate slogan. Keep this up until the company is ready to make the workplace safer.

Warm-Out or Cold-Out

If the workplace is too hot or too cold, come to work, but only stay in those parts of the workplace that are comfortable. Refuse to work where the temperature is dangerous to your health or your pregnancy.

Overkill

If you work in a large plant or factory where many workers are exposed to potential damage, put in writing every single reproductive health hazard complaint that could be turned over to OSHA. For each item, keep a record of when the complaint was submitted, whether or when needed repair materials were ordered, whether or when the hazard-reducing work was ever completed. With the threat of hundreds of documented cases of purported hazard violations ready to be submitted to OSHA for investigation, it may be cheaper and easier for management to alleviate the worst cases.

Be Your Own Reproductive Health Detective

Try to trace the specific source of the problem. You can enlist the support of both male and female workers. Have them write down any part of the work process, location, or time of day when they have suspicions that toxic substances to the reproductive system are accumulating. Have the employees be as detailed as possible, describing odors, vapors, color changes, dust particles, etc., and any immediate or delayed symptoms they suffer.

Contract Language

Try negotiating into your contract the right of safety committee members to "red tag" a machine which is causing an imminent danger. When a machine is red-tagged, it cannot be used until it is satisfactorily fixed and a committee member removes the tag. This clause should not be limited to union contracts in industrial plants and manufacturing industries, but is valuable in almost all occupational settings. For example, in hospitals there are many pieces of equipment that are potentially dangerous. Some emit radiation; others leak anesthetic gases or ethylene oxide if they are not properly maintained and monitored. In schools, there is art and scientific lab equipment and in offices, VDTs and photocopiers that can pose hazards.

HOW TO INFLUENCE YOUR LEGISLATURE

Some states have Women's Bureaus. These often provide guidance to novices who are trying to influence or initiate legislation beneficial to women. It is frequently easier to achieve success at the state or local level than at the federal level where so many different interest groups from all over the country exert pressure. If you want to introduce a bill to improve the conditions of the pregnant worker, it is a good idea to get a powerful legislator to introduce the bill. These powerful lawmakers are usually owed many favors and the bill has a greater chance of passage.

The first and most important step toward success is to know your legislators very well. This means getting to know the staff members as well as the lawmakers themselves. Staff members are crucial as they arrange the appointments and calendar and bring items to the lawmaker's attention. Try to personalize reproductive hazards of the workplace. Inform the legislative staff about risks that they may be unaware of. Particularly emphasize risks that their family members and friends might face in their workplaces.

When this is accomplished, your next step is to lobby the legislature. The lawmakers have to know that you can sway a large bloc of voters whose support the legislator must count on for reelection. Reproduction is of concern to everyone, not just the potential parents. Grandparents want to have healthy grandchildren and may not know that this desire can be thwarted by toxic substances in the workplace. Tie your legislative request to as wide a range of voter interest as possible.

Lobbying is tedious, time-consuming, and requires a great deal of political savvy. Officials from Women's Bureaus have reported that women who came to them for legislative advice were frequently uneasy about lobbying – feeling it wasn't a nice thing to do. Luckily, these "ladylike" qualms are fast disappearing.

It is also essential to be extremely well informed about issues. You need to know more than anyone else about the plight of pregnant workers and you must be honest or you will not be trusted.

You or your representative must become politically astute, being tuned in to the political climate – what is possible, how to present the bill, and what interest of the legislator you are going to tap.

Find out which groups you can coalesce with – both natural allies and others that might have just this one issue in common with you. Coalitions usually have much more clout than single group endeavors but are intrinsically fragile and vulnerable. Aside from the one issue of agreement, groups may be dissimilar in other areas important to the membership. Both Right-to-Life and Pro-Choice groups would probably support legislation reducing substances hazardous to reproductive health in the workplace, but with regard to abortion they would be enemies rather than friends.

If you do decide that an "unholy alliance" is necessary to achieve your goal, remember to settle differences in private, whatever the provocation. If you go public with your differences, your coalition will disintegrate.

Latch onto and harness the concerns of the electorate even though they might not be the same as yours. For example, you are concerned with the direct risk to your pregnancy and the electorate is concerned about long-term economic and health costs. After the Union Carbide accident in Bhopal, India, communities became much more concerned about the health costs associated with the industrial plants in their midst.

You must also learn who your opponents are and their strengths and weaknesses. Otherwise your efforts will be torpedoed early in the fight.

Lobbying involves many details, such as insuring that you get on the mailing list of the appropriate legislative committee so that you know when there is a public hearing on a crucial bill. You need to move very fast in order to have your position represented and represented well. It is, therefore, useful to have written and oral testimony and presenters ready to go.

Not only is it essential to promote existing legislation to protect your interests, it is equally vital to kill a pending bill if it is harmful to your welfare as a pregnant worker. Try to get the bill killed in committee. It is harder to kill a bill once it has reached the floor.

If you are initiating or promoting a specific bill, get the media interested in your campaign, particularly in the districts in which key legislators live. Media coverage can make or break your campaign. Know the ideological and political perspective of the newspapers, magazines, radio and TV stations. If they have vested interests in certain political or moral positions or have economic connections with corporations that your bill might hurt, they may give you inadequate coverage or even worse, negative coverage.

SUGGESTED READINGS

A Better Place to Work, 1986, pamphlet published by the Communication Workers of America, Washington, DC.

American Labor, 1981, "Do-It-Yourself Tactics: Local Action on Job Safety," No. 15, American Labor Education Center, Washington, DC.

Kaye, L., 1986, *Reproductive Hazards in the Workplace: Some Case Studies*, Lynn Kaye, National Action Committee on the Status of Women, Toronto, Ontario, Canada.

ACORN, *Community Organizing Handbook No. 3: Actions and campaigns*, Publications Director, ACORN, 401 Howard Ave., New Orleans, LA 70130 ($2.50 plus 10% postage and handling charges); an overview of the elements of successful actions and campaigns, including 5 case studies.

Moore, A. O., *A Citizen's Guide to Legal Action and Organizing*, Environmental Action Foundation, 1525 New Hampshire Ave. N.W. Washington, DC 20036 ($15 plus 10% for postage and handling); a workbook for toxics victims.

Staples, L., *Roots to Power*, Praeger, NY ($12.95, paperback); a nuts and bolts primer on how to organize using the ACORN model.

Appendix A

Protective Legislation and Governmental Regulatory Agencies

PROTECTIVE LEGISLATION

1910: New York state was first to enact a state Workers' Compensation Law. It took until 1963 for all 50 states to enact such legislation. These laws attempted to provide comparatively rapid and fair compensation for workplace-induced accidents and illnesses. The costs to employers of insurance and payments to injured workers were intended to act as deterrents against unhealthy and unsafe workplaces. Because workers could not sue their employers for injuries under the law, the hoped-for deterrent effect was not achieved.

> Workers usually cannot meet the criteria for eligibility in cases of occupationally-induced reproductive harm. These criteria refer to a personal injury or disease (not to a fetus or spouse) resulting in a job disability and caused by a workplace accident or exposure, (hard to prove). Most cases of reproductive damage, e.g., sterility due to toxic exposure, do not impair one's ability to work. Therefore, in such cases there would probably be no way to recover from the employer. The compensation system is premised on paying for only those losses which handicap the worker's capability to compete for a job.

1962: The Kennedy administration outlawed discrimination in the federal civil service.

1963: Congress passed an Equal Pay Act prohibiting different pay for men and women working at equivalent jobs. This act did not deal with equal opportunity, excluded domestics and farm laborers, and did not forbid employers from refusing to employ women.

1964: Congress passed the Civil Rights Act. Sex became part of a clause in Title VII of the Act that prohibited firms with 15 or more employees and labor unions with 15 or more members from discriminating on account of religion, race, and ethnicity. This law provides the strongest protection for women workers by prohibiting sex discrimination in hiring, layoffs, promotions, training, seniority, and discharge practices.

1964: The Equal Employment Opportunity Commission was created to administer Title VII. Its job was to furnish guidelines for employers to follow in order to avoid sex discrimination charges.

1968: Executive Order 11375 forbade federal contractors from discriminating against women. It required the contractors to file affirmative action programs indicating how they planned to improve job opportunities for women and minorities.

1970: Congress passed the Occupational Safety and Health Act which created the Occupational Safety and Health Administration (OSHA) within the Department of Labor. OSHA sets and enforces health and safety standards governing the work environment.

1972: By this date, all state legislatures had passed bills prohibiting sex discrimination in employment.

1978: Congress passed the Pregnancy Disability Amendment to Title VII of the 1964 Civil Rights Act. Discrimination against pregnant women was considered to be sexual discrimination. Pregnancy and maternity were defined as temporary disabilities. If companies carried disability plans, pregnant women would receive the same income and/or medical benefits granted to any other temporarily disabled employee. This would usually cover the period from 4 to 6 weeks before delivery to 4 to 6 weeks after delivery.

1983: Congress passed the Federal Hazards Communication Standard (FHCS) which covered 14 million workers in 300,000 manufacturing plants. In 1987, the standard was expanded to cover 81 million workers in 5 million worksites. Twenty-four states and numerous municipalities also passed Right-to-Know (RTK) Laws. The purpose of these laws was to make sure that workers knew the hazards of the substances with which they worked.

> Under the present interpretation of the law, OSHA claims that the federal standard preempts (overrides) state and municipal Right-to-Know laws except those aspects pertaining to community information requirements and coverage of state public employees. The 1987 standard is stronger than the earlier one and enforcement is likely to be a more important factor than coverage.
>
> Physical hazards such as noise, radiation, heat and cold, vibration, repetitive trauma, and safety hazards were not covered by the Hazard Communications Standard. More pregnant workers are subjected to physical hazards than they are to chemical toxics. The Alaska RTK law was recently expanded to cover physical hazards. Additional states are expected to follow.

GOVERNMENTAL REGULATORY AGENCIES

1. The Occupational Safety and Health Administration (OSHA)

OSHA was created as a regulatory agency within the Department of Labor and is mandated to enforce the provisions of the 1970 Occupational Safety and Health Act. It sets and enforces standards, conducts inspections of workplaces, insures that employers maintain records of their health and safety experiences, and investigates complaints. Workers can file complaints and seek protection under the General Duty Clause of the Act. This clause states that an employer "shall furnish to each of his employees employment and a place of employment which are free from recognized hazards that are causing or are likely to cause death or serious physical harm to his employees."

2. The National Institute for Occupational Safety and Health (NIOSH)

NIOSH is part of the Centers for Disease Control (CDC) of the U.S. Public Health Service. It is the research counterpart of OSHA. NIOSH conducts research on occupational diseases and makes recommendations for the standards OSHA should use and investigates methods to improve technologies to measure and control toxic emissions.

3. The Environmental Protection Agency (EPA)

The Toxic Substance Control Act (TSCA) gives the EPA power to control risks to the environment and human health caused by the production, use, and disposal of toxic substances. The EPA also regulates certain occupational exposures to reproductive and other health hazards. It has jurisdiction over pesticide and herbicide exposure to farm workers and radiation exposure to employees.

4. The Nuclear Regulatory Commission (NRC)

The Energy Reorganization Act of 1974 created the Nuclear Regulatory Commission (NRC) which has the primary authority for the regulation of occupational exposure to ionizing radiation in the nuclear industry. It is supposed to use its licensing procedure to insure safety. However, it only establishes minimum criteria.

Appendix B

Alphabetical List of Resource Organizations

American Labor Education Center
1835 Kilbourne Pl., NW
Washington, DC 20010
(202) 387-6780

ACOSH/Industrial Hygiene
1435 N. Fremont Ave.
Room 128
Tucson, AZ 85721
(602) 626-1659

AFL-CIO
Health and Safety Department
815 16th St.
Washington, DC 20006
(202) 637-5000

Akwesasne Environment/Mothers' Milk Project
c/o Katsi Cook
226 Blackman Hill Road
Berkshire, NY 13736
(607) 657-8413

Alabama Department of Public Health
Epidemiology Division
900 State Office Building
Montgomery, AL 36130
(205) 261-5130

Alaska Department of Labor
Occupational Safety and Health Section
Box 1149
Juneau, AK 99802
(907) 465-4856

ALCOSH
100 East Second Street
Jamestown, NY 14701
(716) 488-0720

Alice Hamilton Occupational Health Center
410 Seventh St., SE
Washington, DC 20003
(202) 543-0005

Amalgamated Clothing and Textile Workers Union
Health & Safety Department
15 Union Square West
New York, NY 10003
(212) 242-0700

American Academy of Environmental Medicine
P.O. Box 16106
Denver, CO 80216

American Federation of State, County and Municipal Employees
Health and Safety Department
1625 L Street, NW
Washington, DC 20036
(202) 452-4800

American Labor Education Center
1835 Kilbourne Pl., NW
Washington, DC 20010
(202) 387-6780

American Lung Association
1740 Avenue of the Americas
New York, NY 10020
(212) 315-8700

American Trial Lawyers Association (ATLA)
1050 31st St., NW
Washington, DC 20007
(202) 965-3500

Arizona Industrial Commission
Division of Occupational Safety and Health
800 West Washington St.
P.O. Box 19070
Phoenix, AZ 85007
(602) 255-5795

Arkansas Department of Labor
Safety-Industrial Hygiene
616 Garrison, Room 209
Fort Smith, AR 72901
(501) 783-2103

Arkansas Reproductive Health Monitoring System
University of Arkansas
Medical Science
804 Wolfe St.
Little Rock, AR 72202-3591
(501) 375-3925

Asbestos Victims Special Trust Fund
1500 Walnut Street
Mezzanine Floor
Philadelphia, PA 19102
(215) 735-1188

Association of Occupational & Environmental Clinics
1030 15th St., NW
Suite 410
Washington, DC 20005
(202) 682-1807

Association of Birth Defect Children
3526 Emerywood Lane
Orlando, FL 32812
(507) 629-1466

BACOSH
Labor Occupational Health Program
Institute of Industrial Relations
2521 Channing Way
Berkeley, CA 94720
(415) 642-5507

Barlow Hospital
Occupational Medicine Department
2000 Stadium Way
Los Angeles, CA 90026-2696
(213) 250-4200

Bassett Farm Safety and Health Project
Mary Imogene Bassett Hospital
Cooperstown, NY 13326
(607) 547-3971

Brigham & Women's Hospital
Occupational & Environmental Health Clinic
75 Francis St.
Boston, MA 02115
(617) 332-7257

Brown and White Lung Association
1209 Pendleton Street
Greenville, SC 29602
(803) 269-8048

California Department of Health Services
Health Hazard Surveillance & Evaluation Services
2151 Berkeley Way
Berkeley, CA 94704
(415) 540-3141

California Department of Industrial Relations
Division of Occupational Safety and Health
525 Golden Gate Avenue
San Francisco, CA 94102
(415) 557-1946

California Teratogen Registry
University of California at San Diego H-814-B
225 W. Dickenson Street
San Diego, CA 92103
(619) 294-3584

Cambridge Hospital
Occupational Health Program & Environmental Health Center
1493 Cambridge Street
Cambridge, MA 02139
(617) 498-1580

Center for Occupational & Environmental Health
Johns Hopkins University
Wyman Park Health Service
Building 6, 3100 Wyman Park Dr.
Baltimore, MD 27211
(301) 338-3376

Center for Safety in the Arts
5 Beekman Place
New York, NY 10038
(212) 227-6220

Central New York Occupational Health Clinical Center (CNYOHCC)
c/o SUNY Health Sciences Center
750 East Adams St.
Syracuse, NY 13210
(315) 473-5466

Central New York Committee on Occupational Safety and Health (CNYCOSH)
615 W. Genesee Street
Syracuse, NY 13204
(315) 471-6187

Chemical Industry Institute of Toxicology
P.O. Box 12137
Research Triangle Pk., NC 27709
(919) 541-2070

Chicago Area Committee on Occupational Safety and Health (CACOSH)
37 S. Ashland Street
Chicago, IL 60607
(312) 666-1611

Coal Employment Project & Coal Mining Women's Support Team
P.O. Box 3403
Oak Ridge, TN 37830
(615) 482-3428

Coalition of Labor Union Women (CLUW)
15 Union Square West
New York, NY 10003
(212) 242-0700

Coalition on Environmental and Occupational Health Hazards
Toxics Coordinating Program
2609 Capitol Ave.
Sacramento, CA 95816
(916) 441-4077

Colorado Department of Health Epidemiology and Disease Control Division
4210 East 11th Ave.
Denver, CO 80220
(303) 320-8333

Communications Workers of America
Health & Safety Department
1925 K St., NW
Washington, DC 20006
(202) 728-2483

Congressional Caucus for Women's Issues
2471 Rayburn HOB
Washington, DC 20515
(202) 387-4709

ConnectiCOSH
130 Huyhope Street
P.O. Box 31107
Hartford, CT 06106
(203) 549-1877

Connecticut Labor Department
Occupational Safety and Health Division
200 Folly Brook Boulevard
Wethersfield, CT 06109
(203) 566-4384

Cook County Hospital
Division of Occupational Medicine
20 S. Walcott, 13th Fl.
Chicago, IL 60612
(312) 633-5310

Cornell University
New York State School of Industrial & Labor Relations
Chemical Hazard Information Program
120 Delaware Ave.
Buffalo, NY 14202
(716) 842-1124

CUNY Center for Occupational and Environmental Health
425 E. 25th Street
New York, NY 10010
(212) 481-4361

Delaware Department of Labor
Division of Industrial Affairs
Occupational Safety & Health Section
820 North French Street
Wilmington, DE 19801
(302) 571-2879

District of Columbia
Department of Employment Services
Office of Occupational Safety and Health
500 C Street, NW
Washington, DC 20001
(202) 639-1000

Ecological Illness Law Report
Earon Davis, JD, MPH, editor
P.O. Box 6099
Wilmette, IL 60091
(312) 256-3730

ENYCOSH
IUE Local 301
121 Erie Blvd.
Schenectady, NY 12305
(518) 438-5003

Farmworker Justice Fund
Occupational Health Project
2001 S St., NW, Suite 210
Washington, DC 20009
(202) 462-8192

Florida Department of Labor and Employment Security
Bureau of Industrial Safety and Health
204 Lafayette Building
Tallahassee, FL 32399-2152
(904) 488-3044

Francis Scott Key Medical Center
4940 Eastern Avenue
Baltimore, MD 21224
(301) 955-0537

George Washington University
School of Medicine
Occupational Medicine Program
2150 Pennsylvania Ave.
Washington, DC 20037
(202) 676-5366

Georgia Department of Human Resources
Occupational Health
47 Trinity Avenue, SW
Atlanta, GA 30334
(404) 894-6644

Good Samaritan Hospital
Occupational Health
3217 Clifton Ave.
Cincinnati, OH 45220-2489
(513) 872-2875

Greater Cincinnati Occupational Health Center
2450 Kipling Ave.
Suite 203
Cincinnati, OH 45239
(513) 541-0561

Hawaii Department of Labor and Industrial Relations
Division of Occupational Safety and Health
830 Punchbowl Street
Honolulu, HI 96813
(808) 548-6465

Health Policy Advisory Center (Health/PAC)
17 Murray Street
New York, NY 10007
(212) 267-8890

Health/Rights: Collaborative Project on Reproductive Health Hazards
c/o ACLU
132 West 43rd Street
New York, NY 10036
(212) 944-9800

Healthy Mothers/Healthy Babies Coalition
409 12th St., NW
Washington, DC 20024-2188

Highlander Research and Education Center
Rt. 3, Box 370
New Market, TN 37820
(615) 933-3443

Indiana Department of Labor
Indiana Occupational Safety and Health Administration
1013 State Office Building
100 North Senate Ave.
Indianapolis, IN 46204
(317) 232-2693

Institute for Women & Work
Cornell University School for Industrial and Labor Relations
15 East 26th St., 4th Floor
New York, NY 10010
(212) 340-2823

Iowa Department of Employment Services
Occupational Safety and Health Enforcement Bureau
1000 East Grand Avenue
Des Moines, IA 50319
(515) 281-3445

Iowa State Department of Health
Occupational and Environmental Hazards Division
Lucas State Office Building
Des Moines, IA 50319
(515) 281-4928 and -7782

Jobmed
St. Mary of Nazareth Hospital
2233 W. Division
Chicago, IL 60622
(312) 770-3275

JOHCOSH
CNY Labor Agency
Mayro Building, Room 128
239 Genesee St.
Utica, NY 13501
(315) 735-6101

Johns Hopkins University
Center for Occupational & Environmental Health
Building 6, 3100 Wyman Pk. Dr.
Baltimore, MD 21210
(301) 338-3501

Kansas Department of Human Resources
Division of Industrial Safety and Health
401 S.W. Topeka Boulevard
Topeka, KS 66603
(913) 296-4386

Kentucky Department of Labor
Occupational Safety & Health Program
U.S. 127 South
Frankfort, KY 40601
(502) 564-6895

Kentucky Labor Cabinet
Division of Occupational Safety & Health Compliance
U.S. 127 South Building
Frankfort, KY 40601
(502) 564-3070

Labor Education and Research Project
P.O. Box 20001
Detroit, MI 48220
(313) 883-5580

Labor Education Service
Women's Programming
University of Minnesota
437 Management/Economics Building
271 19th Ave., South
Minneapolis, MN 55455
(612) 624-5020

Labor Institute
853 Broadway
New York, NY 10003
(212) 674-3322

LACOSH
2501 S. Hill Street
Los Angeles, CA 90007
(213) 749-6161

League of Women Voters
Project on Toxics and Community Health
1730 M St., NW
Washington, DC 20035
(202) 429-1965

Louisiana Department of Labor
Division of Occupational Safety and Health
P.O. Box 94094
Baton Rouge, LA 70804
(504) 342-3011

Loyola University
Institute of Human Relations
Labor Studies Program
Box 12
New Orleans, LA 70118
(504) 861-5830

Maine Department of Labor
Division of Industrial Safety
State House, Station #45
Augusta, ME 04333
(207) 289-3788

Maine Labor Group on Health, Inc.
Box V
Augusta, ME 04330
(207) 622-7823

March of Dimes Birth Defects Foundation
1275 Mamaroneck Ave.
White Plains, NY 10605
(914) 428-7100

Maryland Division of Labor and Industry
Occupational Safety and Health Program
501 St. Paul Place
Baltimore, MD 21202
(301) 333-4195

Maryland Occupational Safety and Health Administration
501 St. Paul Pl., 3rd Fl.
Baltimore, MD 21202
(301) 333-4133

Massachusetts Department of Labor and Industries
Occupational Hygiene
100 Cambridge St., 11th Fl.
Boston, MA 02108
(617) 727-3454

Massachusetts Public Interest Research Group (MASSPIRG)
29 Temple Place
Boston, MA 02111
(617) 292-4800

MassCOSH
555 Amory St.
Boston, MA 02115
(617) 524-6686

Medical College of Pennsylvania
Division of Occupational Health
3300 Henry Ave.
Philadelphia, PA 19129
(215) 842-6540

Meharry Medical College
Occupational Medicine Clinic
Department of Community & Occupational Health
1005 D.H. Todd Blvd.
Nashville, TN 37208
(615) 327-6734

Michigan Department of Public Health
Division of Environmental & Occupational Health
3500 N. Logan Street
Box 30035
Lansing, MI 48909
(517) 335-8250

Michigan Department of Labor
MIOSHA Information Division
Bureau of Safety & Regulations
7150 Harris Drive
Box 30015
Lansing, MI 48909
(517) 322-1851

Minnesota Department of Labor and Industry
Occupational Safety and Health
St. Paul, MN 55101
(612) 296-2342

Mississippi State Department of Health
Occupational Safety & Health
2423 North State Street
P.O. Box 1700
Jackson, MS 39215
(601) 982-6315

Missouri Department of Labor and Industrial Relations
Division of Labor Standards
621 East McCarty
P.O. Box 449
Jefferson City, MO 65102
(314) 751-3403

Montana State Department of Health
& Environmental Services
Cogawith Building
Helene, MT 59620
(406) 444-2544

Mount Sinai School of Medicine
Occupational Health Clinic
1 Gustave L. Levy Place
New York, NY 10029
(212) 241-6173

National Cancer Institute
Cancer Information Clearinghouse
Office of Cancer Communications
7910 Woodmont Avenue
Bethesda, MD 20205
(301) 496-4000

National Capital Lactation Center
3800 Reservoir Road, NW
Washington, DC 20007
(202) 625-MILK

National Center for Environmental Health Strategies
1100 Rural Avenue
Voorhees, NJ 08043
(609) 429-5358

National Center for Policy Alternatives
Children & Toxics
2000 Florida Ave., NW
Washington, DC 20009
(202) 387-6030

National Coalition Against the Misuse of Pesticides
530 7th St., SE
Washington, DC 20003
(202) 543-5450

National Coalition of Injured Workers
c/o Injured Workers of RI
340 Lockwood Street
Providence, RI 02907
(401) 828-6520

National Ecological & Environmental Delivery System (NEEDS)
602 Nottingham Road
Syracuse, NY 13224

National Foundation for the Chemically Hypersensitive
P.O. Box 9
Wrightsville, NC 28480
(919) 256-5391

National Institute for Occupational Safety & Health
4676 Columbia Parkway
Cincinnati, OH 45226
(513) 533-8236

National Jewish Center for Immunology and Respiratory Medicine
Occupational Medicine
1400 Jackson St.
Denver, CO 80206
(303) 398-1526

National Network to Prevent Birth Defects
Box 15309, SE Station
Washington, DC 20003
(202) 543-5450

National Safe Workplace Institute
122 S. Michigan Ave.
Suite 1450
Chicago, IL 60603
(312) 939-0690

National Toxics Campaign
37 Temple Place
Boston, MA 02111
(617) 482-1477

National Tradeswomen Network
c/o Chicago Women in the Trades
37 S. Ashland Ave.
Chicago, IL 60607
(312) 942-0802

National Women's Health Network
1325 G St., NW
Lower Level B
Washington, DC 20005
(202) 347-1140

National Women's Law Center
1616 P St., NW
Washington, DC 20036
(202) 328-5160

Nebraska State Department of Health
P.O. Box 95007
Lincoln, NB 68509
(402) 471-2937

Nevada State Labor Commission
Division of Occupational Safety and Health
Capitol Complex
Carson City, NV 89710
(702) 885-3250

New Hampshire Division of Public Health Services
6 Hazen Drive
Concord, NH 03301
(603) 271-4501

New Jersey State Department of Health
Environmental Health
CN 360
Trenton, NJ 08625
(609) 633-2043

New Jersey Work Environment Council
452 E. Third St.
Morristown, NJ 08057
(609) 886-9405

New Mexico Health and Environment Department
Occupational Health & Safety
P.O. Box 968
Santa Fe, NM 87504
(505) 827-2879

New Mexico Public Interest Research Group
Box 66, Room 96
Student Union Building, UNM
Albuquerque, NM 87131
(505) 277-2758

New York Committee on Occupational Safety and Health (NYCOSH)
275 Seventh Ave., 8th Fl.
New York, NY 10001
(212) 627-3900

New York State Department of Health
Division of Occupational & Environmental Health
Empire State Plaza
Albany, NY 12237
(518) 458-6202

New York State Department of Labor
Division of Safety and Health
State Campus
Albany, NY 12240
(518) 457-3518

Nine to Five National Association of Working Women
614 Superior Ave., NW
Cleveland, OH 44113
(216) 566-9308

NIOSH Appalachian Laboratory for Occupational Safety & Health
944 Chestnut Ridge Rd.
Morgantown, WV 26805
(304) 599-7521

Norfolk County Hospital
Occupational Health Services
2001 Washington Street
Braintree, MA 02184
(617) 843-0690

North Carolina Department of Labor
Occupational Health and Safety Division
4 West Edenton Street
Raleigh, NC 27601
(919) 733-2385

North Carolina Division of Health Services
Occupational Health Branch
P.O. Box 2091
Raleigh, NC 27602
(919) 733-3680

North Carolina Occupational Safety & Health Project (NCOSH)
P.O. Box 2514
Durham, NC 27705
(919) 286-9249

North Dakota State Department of Health
Capitol Building
Bismarck, ND 58505
(701) 224-2372

Occupational Health Center
1001 East Palmdale
Tucson, AZ 85714
(602) 889-9574

Occupational Medical Centers, Inc.
490 L'Enfant Plaza E., SW
Suite 4300
Washington, DC 20024
(202) 488-7990

Occupational & Environmental Reproductive Hazards
Clinic & Education Center
University of Massachusetts Medical Center
55 Lake Ave., North
Worcester, MA 01655
(508) 856-2818

Occupational Health Foundation Workers' Institute for Safety and Health
1126 16th St., NW, Suite 413
Washington, DC 20036
(202) 887-1980

Occupational Health and Safety Bureau
P.O. Box 968
Santa Fe, NM 87501
(505) 827-2877

Occupational Medicine Clinic
c/o Dept. of Community & Preventive Medicine
SUNY, Stony Brook
Stony Brook, NY 11794
(516) 444-2460

Occupational Safety and Health Law Center
1536 16th St., NW
Washington, DC 20036
(202) 328-8300

Office Technology Education Project
241 St. Botolph St.
Boston, MA 02115
(617) 526-8324

Office of Technology Assessment
Congress of the United States
Washington, DC 20510
(202) 226-2115

Ohio Department of Health
Occupational Health
246 North High Street
P.O. Box 118
Columbus, OH 43216
(614) 466-4183

Oil, Chemical and Atomic Workers Union (OCAW)
Health & Safety Department
P.O. Box 2812
Denver, CO 80201
(303) 987-2229

Oklahoma Department of Labor
Consultation Division, Occupational Safety and Health
1315 Broadway Place
Oklahoma City, OK 73103
(405) 235-0530

Oklahoma State Health Department
1000 NE 10th Street
Oklahoma City, OK 73152
(405) 271-5420

Ontario Workers' Health Centre
1292 Barton St., East
Hamilton, Ontario, CANADA L8H 2W1
(416) 544-1561

Oregon Health Sciences University
Program on Occupational Diseases
3181 S W Jackson Park Rd.
Portland, OR 97201
(503) 220-5742

Pennsylvania Department of Labor and Industry
Bureau of Occupational and Industrial Safety
7th & Forster Streets
Harrisburg, PA 17120
(717) 787-3323

Philadelphia Area Project on Safety and Health (PhilaPOSH)
3001 Walnut St., 5th Floor
Philadelphia, PA 19104
(215) 386-7000

PIRG National Toxics Action Campaign
29 Temple Pl.
Boston, MA 02111
(617) 292-4800

Public Citizen Health Research Group
2000 P St., NW
Washington, DC 20036
(202) 872-0320

Puerto Rico Department of Labor
Occupational Safety & Health
505 Munoz Rivera Ave.
Hato Rey, PR 00918
(809) 754-2172

Reproductive Toxicology Center
2425 L Street, NW
Washington, DC 20037
(202) 293-5137

Rhode Island Committee on Occupational Safety & Health (RICOSH)
340 Lockwood Street
Providence, RI 02907
(401) 751-2015

Rhode Island Department of Health
Radiation/Occupational Health
75 Davis Street
Providence, RI 02908
(401) 277-2438

Rhode Island Department of Labor
Division of Occupational Safety
220 Elmwood Avenue
Providence, RI 02907
(401) 457-1843

Rochester Committee on Occupational Safety & Health (ROCOSH)
502 Lyell Ave., Suite #1
Rochester, NY 14606
(716) 458-8553

Rochester Regional Occupational Health Clinic
c/o University of Rochester
Medical Center
601 Elmwood Ave., Box 644
Rochester, NY 14642
(716) 275-2191

Rocky Mountain Center for Occupational & Environmental Health
Building 512
University of Utah Medical Center
Salt Lake City, UT 84112
(801) 581-4800

Sacramento COSH (SACOSH)
c/o Fire Fighters Local 522
3101 Stockton Boulevard
Sacramento, CA 95820
(916) 444-8134

San Francisco General Hospital
Occupational Health Clinic
Building 9, Room 109
2550 23rd Street
San Francisco, CA 94110
(415) 821-5391

Santa Clara COSH (SCCOSH)
760 N. 1st Street
San Jose, CA 95112
(408) 998-4050

Service Employees International Union
Health & Safety Department
1313 L Street, NW
Washington, DC 20005
(202) 898-3200

Silicon Valley Toxics Coalition
277 W. Hedding St., #208
San Jose, CA 95110
(408) 287-6707

Society for Occupational & Environmental Health
2021 K St., NW, Suite 305
P.O. Box 42360
Washington, DC 20015-0360

Society of Toxicology
1133 15th St., NW
Washington, DC 20005

South Carolina Department of Health and Environmental Control
Health Hazard Evaluation
2600 Bull St.
Columbia, SC 29201
(803) 734-5429

South Carolina Department of Labor
3600 Forest Drive, 4th Fl.
Columbia, SC 29204
(803) 734-9600

South Dakota State Department of Health
Division of Public Health
Joe Foss Building
523 E. Capitol Ave.
Pierre, SD 57501
(605) 773-3364

South East Michigan Committee on Occupational Safety & Health (SEMCOSH)
2727 Second Street
Detroit, MI
(313) 961-5685

Southern Occupational Health Center
UCLA School of Public Health
Los Angeles, CA 90024
(213) 825-7066

State of Wyoming Environmental Health
Hathaway Building, Room 482
Cheyenne, WY 82002
(307) 777-7957

TENNCOSH
c/o Center for Health Services
Station 17
Vanderbilt Medical Center
Nashville, TN 37235
(615) 322-4773

Tennessee Department of Labor
Occupational Safety and Health
501 Union Building
Nashville, TN 37219
(615) 741-2582

Texas Department of Health
1100 West 49th Street
Austin, TX 78756-3192
(512) 458-7497

TEXCOSH
c/o OCAW Local 4243
5735 Regina
Beaumont, TX 77706

Trial Lawyers for Public Justice
1625 Massachusetts Ave., NW
Suite 100
Washington, DC 20036
(202) 797-8600

U.S. Department of Labor
OSHA Region I Office
JFK Federal Building
Government Center
Boston, MA 02203
(617) 565-7159

U.S. Department of Labor
OSHA Region II office
201 Varick Street
New York, NY 10014
(212) 337-2378

U.S. Department of Labor
OSHA Region III Office
Gateway Building, Suite 2100
3535 Market Street
Philadelphia, PA 19104
(215) 596-1201

U.S. Department of Labor
OSHA Region IV Office
1375 Peachtree St., NE
Suite 587
Atlanta, GA 30367
(404) 347-3573

U.S. Department of Labor
OSHA Region V Office
230 S. Dearborn, Room 3244
Chicago, IL 60644
(312) 353-2220

U.S. Department of Labor
OSHA Region VI Office
525 Griffen Street
Dallas, TX 75202
(214) 767-4731

U.S. Department of Labor
OSHA Region VII Office
911 Walnut St.
Kansas City, MO 64106
(816) 374-5861

U.S. Department of Labor
OSHA Region VIII Office
Federal Building, Room 1554
1961 Stout St., #294
Denver, CO 80294
(303) 844-3061

U.S. Department of Labor
OSHA Region IX Office
450 Golden Gate Avenue
San Francisco, CA 94102
(415) 556-7260

U.S. Department of Labor
OSHA Region X Office
Federal Office Building
909 First Ave., Room 6003
Seattle, WA 98174
(206) 442-5930

U.S. Department of Labor
Technical Data Center
OSHA National Office
200 Constitution Ave., NW
Washington, DC 20210
(202) 523-9700

U.S. Environmental Protection Agency
410 M St., SW
Washington, DC 20460
(800) 555-1404

U.S. Nuclear Regulatory Commission
Office of Public Affairs
Washington, DC 20553
(301) 492-7715

Union Occupational Health Center
450 Grider Street
Buffalo, NY 14215
(716) 894-9366

United Automobile, Aerospace & Agricultural Workers Union
Health & Safety Department
8000 E. Jefferson Ave.
Detroit, MI 48214
(313) 926-5000

United Farm Workers
P.O. Box 62
Keene, CA 93531
(805) 822-5571

United Food & Commercial Workers
Suffridge Building
1775 K St., NW
Washington, DC 20006
(202) 223-3111

United Mine Workers of America
Health & Safety Department
900 15th St., NW
Washington, DC 20005
(202) 842-7207

United Steelworkers of America
Health & Safety Department
5 Gateway Center
Pittsburgh, PA 15222
(412) 562-2580

University of Arizona
College of Medicine
Environmental, Preventive & Occupational Health Clinic
1450 N. Cherry St.
Tucson, AZ 85724
(602) 626-7900

University of California-Berkeley
School of Public Health
Northern California Occupational Health Center
Berkeley, CA 94720
(415) 642-1681

University of California-Davis
Occupational Health Clinic
2315 Stockton Blvd.
Trailer #1530
Sacramento, CA 95817
(916) 453-5234

University of California-Los Angeles
UCLA Labor Center
Occupational Safety & Health Program
1001 Gayley Ave.
Los Angeles, CA 90024
(213) 825-7012

University of Connecticut
Project on Women and Technology
417 Whitney Rd., U-118
Storrs, CT 06268
(203) 486-4738

University of Illinois-Chicago
School of Public Health
Environmental & Occupational Health Sciences
P.O. Box 6998
Chicago, IL 60680
(312) 996-8855

University of Illinois
College of Medicine
Great Lakes Center for Occupational Health
P.O. Box 6998
Chicago, IL 60680
(312) 996-7887

University of Massachusetts
Occupational & Environmental Reproductive Hazards Clinic
& Resource Center
55 Lake Ave., North
Worcester, MA 01655
(508) 856-2818

University of Michigan
School of Public Health
Department of Environmental & Industrial Health
Occupational Health Program
Ann Arbor, MI 48109
(313) 764-2594

University of Minnesota
School of Public Health
Occupational Health Program
Box 197, Mayo Building
420 Delaware St., SE
Minneapolis, MN 55455

University of New Mexico
Occupational Health Program
Family Practice/Psychology Building
Albuquerque, NM 87131
(505) 277-3253

University of Pittsburgh
Graduate School of Public Health
Departments of Industrial Environmental Health Services
Pittsburgh, PA 15260
(412) 624-3047

University of Pittsburgh
School of Medicine
Occupational & Environmental Health Program
190 Lothrop Street #149
Pittsburgh, PA 15261
(412) 648-3240

University of Texas
Medical Branch-Division of Environmental Toxicology
Ewing Hall
Galveston, TX 77550
(409) 761-1803

University of Washington
Harborview Medical Center
Occupational Medicine Program
325 9th Ave., ZA-66
Seattle, WA 98104
(206) 223-3005

Utah Industrial Commission
Occupational Safety and Health Division
P.O. Box 45580
Salt Lake City, UT 84145
(801) 530-6900

Utah State University
Cooperative Extension Safety Program
UMC 8310
Logan, UT 84322
(801) 750-2760

VANCOSH
616 E. 10th Ave.
Vancouver, BC, CANADA V5T2A5
(519) 254-4192

VDT Coalition c/o Labor Occupational Health Program
2521 Channing Way
Berkeley, CA 94720
(415) 642-5507

Vermont Department of Health
Epidemiology
60 Main Street P.O. Box 70
Burlington, VT 05401
(802) 863-7240

Vermont State Department of Health
Division of Radiological and Occupational Health
10 Baldwin Street
Montpelier, VT 05602
(802) 828-2886

Victims of Fiberglass
11143 Lakeshore North
Lake of the Pines
Auburn, CA 95603
(916) 268-0480

Virginia Department of Labor and Industry
Division of Occupational Safety and Health
205 N. Fourth St.
P.O. Box 12064
Richmond, VA 23241
(804) 786-2391

Western MassCOSH
458 Bridge Street
Springfield, MA 01103
(413) 247-9413

Western Washington Toxics Coalitions
4512 University Way, NE
Seattle, WA 98105
(206) 632-1545

White Lung Association
901 Broad Street
Newark, NJ 07012

Wisconsin Division of Health
Section on Occupational Health, Room 112
1414 E. Washington Ave.
Madison, WI 53703
(608) 266-9379

WISCOSH
1334 South 11th St.
Milwaukee, WI 53204
(414) 643-0928

WNYCOSH
450 Grider Street
Buffalo, NY 14215
(716) 897-2110

Women in the Building Trades
555 Amory Street
Boston, MA 02130
(607) 524-3010

Women's Advocacy Bureau
Louisiana Department of Health & Human Resources
200 Riverside Mall
Baton Rouge, LA 70801
(504) 342-2715

Women's Bureau Region I
U.S. Department of Labor
JFK Federal Building
Room 1600
Boston, MA 02203
(617) 565-1988

Women's Bureau Region II
U.S. Department of Labor
201 Varick Street
New York, NY 10014
(212) 337-2390

Women's Bureau Region III
U.S. Department of Labor
Gateway Building
3535 Market St., Room 13280
Philadelphia, PA 19104
(215) 596-1183

Women's Bureau Region IV
U.S. Department of Labor
1371 Peachtree St., NE
Rm. 323
Atlanta, GA 30367
(404) 347-4461

Women's Bureau Region V
U.S. Department of Labor
230 S. Dearborn
Chicago, IL 60644
(312) 353-7205

Women's Bureau Region VI
U.S. Department of Labor
525 Griffin Sq. Bldg.
Dallas, TX 75202
(214) 767-4993

Women's Bureau Region VII
U.S. Department of Labor
911 Walnut St.
Kansas City, MO 64106
(816) 426-3856

Women's Bureau Region VIII
U.S. Department of Labor
1961 Stout St., Room 1452
Denver, CO 80924
(303) 844-4138

Women's Bureau Region IX
U.S. Department of Labor
71 Stevenson St.
San Francisco, CA 94105
(415) 744-6580

Women's Bureau Region X
U.S. Department of Labor
111 Third Ave.
Seattle, WA 98174
(206) 553-5286

Women's Health Program
Occupational Health Unit
Massachusetts Department of Public Health
150 Tremont Street
Boston, MA 02111
(617) 727-7222

Women's Occupational Stress Resource Center
264 Valencia St.
San Francisco, CA 94103
(415) 864-2364

Women's Rights Clinic
Washington Square Legal Services
715 Broadway
New York, NY 10003

Women's Rights Litigation
Rutgers School of Law
15 Washington St.
Newark, NJ 07102
(201) 648-5637

Work Environment Program
University of Lowell
One University Avenue
Lowell, MA 01854
(508) 934-3259

Worker's Institute for Safety and Health
1126 16th St., NW
Washington, DC 20006
(202) 887-1980

Workers Education
Local 189
c/o Jim Bollen
44 Hollingsworth St.
Lynn, MA 01902

Workplace Health Fund
815 16th St., NW, Suite 301
Washington, DC 20006
(202) 842-7833

WOSH
1109 Tecumseh Rd., East
Windsor, Ontario, CANADA N8W2T1
(519) 254-4192

Wyoming Occupational Health & Safety Department
604 East 25th St.
Cheyenne, WY 82002
(307) 777-7786

Yale Occupational Medicine Program
333 Cedar Street
New Haven, CT 06510
(203) 785-4197

Glossary of Occupational Health Terms

acute effects—effects that are seen shortly after exposure to a toxic material, usually at a fairly high concentration.

administrative controls—management changes in work procedures aimed at reducing worker exposure to occupational hazards, e.g., rotation of workers from high to low risk work areas.

American Conference of Governmental Industrial Hygienists (ACGIH)—professional organization which recommends exposure limits for toxic substances. These are called threshold limit values (TLVs).

animal studies—studies in which animals (e.g., rats, mice, hamsters) are exposed to chemicals for different periods of time and at different levels of exposure to see whether or not any develop cancer, or other diseases, over their lifetime at a significantly higher rate than a "control" group of animals (a group that has not been exposed).

bacterial studies—because genetic material is the same in all living organisms, one way to look at millions of living organisms at one time is to study those simpler than mammals, such as bacteria.

bias—a condition that might invalidate study findings. Three common kinds of bias are selection, observation, and recall bias.

selection bias—volunteers who participate in a study may not be typical of the population they are supposed to represent.

recall bias—persons who have suffered reproductive harm are likely to recall more detailed and accurate information about prior events than those who have had no reproductive problems.

observation bias—occurs when the researcher is influenced by knowledge of whether a subject in the study is a member of the case or control group.

birth defect—an abnormality in an infant that may be seen at birth, or noticed at some point later in development.

breast milk pollution—toxic substances to human health enter the mother's milk, primarily through the food chain, and are transmitted to the nursing infant.

carcinogen—a substance capable of causing cancer.

case-control design—a study based on a comparison of individuals who are suffering from a disease with a matched group of individuals who do not have the disease.

chromosomes—rod-like structures containing the genes which are found in the cell nuclei.

chronic effects—health problems which appear a relatively long time after a person's *first* exposure.

closed building syndrome—the buildup of indoor pollution in buildings having inadequate ventilation systems which affects the health of workers, e.g., by causing headaches, allergies, eye irritation, breathing difficulties.

cohort design—follows a group of people to determine who comes down with the disease being studied. Researcher looks for characteristics or exposures that differentiate between those who become ill and those who do not.

confounder—a factor that is associated with both the exposure *and* the outcome under study, making it difficult for the researcher to determine whether the effect is due to the exposure or to the confounding factor.

congenital—present at birth.

cytogenetic studies—observation of cells under an electron microscope to detect chromosomal abnormalities that might be a result of exposure to a mutagen.

dose-response assessment—a process that determines the relationship between the magnitude of human exposure and the probability of human health effects.

embryo—from conception through the 12th week of pregnancy.

embryotoxic—a substance that is toxic to the embryo.

engineering control—changing equipment design and processes in order to reduce the amount of hazardous substances to which workers are exposed.

epidemiology—the study of causes, patterns, and distribution of diseases in human populations.

fetal protection policy—an occupational health policy that prevents fertile and/or pregnant women from holding certain jobs that are deemed to pose a hazard to future offspring.

fetus—an unborn child from 12 weeks until birth.

genetic defect—an abnormality in the genetic material of cells (the genes or the chromosomes).

gene—a unit of heredity comprising a segment of DNA.

germ cell—the egg or sperm cell containing reproductive material that determines the characteristics that will be inherited by the young from its parents.

ionizing radiation—X rays, gamma rays, alpha and beta rays that release energy capable of causing ionization of atoms or molecules in radiated tissue.

local effect—the action of a substance that occurs at the point of contact, e.g., a skin rash.

mutagen—a chemical substance or physical agent that can cause mutations (changes) in the genetic material of living cells.

mutation—a change (usually harmful) in the genetic material of a cell. When it occurs in the germ cell, the mutation can be passed on to future generations.

neonatal—affecting or relating to the newly born.

NOEL (No Observed Effects Level)—the maximum level of exposure that appears to produce no harmful effects for most individuals.

non-ionizing radiation—radiation that is lower in energy than X rays, e.g., visible light, infrared radiation, microwaves, and radio waves.

personal protective equipment (PPE)—clothing or equipment, such as gloves, hearing protectors, and gas and dust masks worn by the person and designed to reduce exposure to potentially hazardous substances.

placental barrier—the border to the placenta (the organ connecting the embryo to the mother's uterus). Some toxics that the mother is exposed to can cross this barrier.

power of a study—the ability of the study design and sample size to detect a real association between exposure and outcome.

radiation—when X rays, gamma rays, alpha or beta particles pass through matter.

somatic cells—all the cells of the body, other than the germ cells.

spermatogenesis—the development of sperm cells.

synergistic effect—an interaction between exposure to two or more substances or agents. This combination causes a greater effect than exposure to either by itself.

systemic effect—when the action of the chemical or substance does not occur at the point of contact, but travels through the system and damages another organ, e.g., vinyl chloride can enter through the lungs but can cause cancer of the liver.

teratogen—a substance or agent that interferes with the development of the embryo or fetus during gestation.

threshold limit values—ACGIH guidelines that establish the maximum level of specific chemicals to which workers may be exposed without experiencing a harmful effect.

tort—a wrongful act for which an employer can be judged legally liable.

toxic—causing harm to cells, tissues, and organs.

Bibliography of Useful Occupational Health References

Ahlborg, G., Jr., Bodin, L., and Hogstedt, C., "Heavy Lifting During Pregnancy—A Hazard to the Fetus? A Prospective Study," *Int J Epidemiol*. 1990;142:1241-4.

Auerbach, E. Roberts and Wallerstein, N., *ESL for Action: Problem-Posing at Work/English for the Workplace*, Reading, MA: Addison-Wesley, Co., 1987.

Axelsson, G. and Rylander, R. "Outcome of Pregnancy in Women Engaged in Laboratory Work at a Petrochemical Plant," *Am J Ind Med*. 1989;16:539-45.

Barlow, S.M. and Sullivan, F.M., *Reproductive Hazards of Industrial Chemicals: An Evaluation of Animal and Human Data*, London: Academic Press, 1982.

Bernhardt, J.H., "Potential Workplace Hazards to Reproductive Health. Information for Primary Prevention," *J Obstet Gynecol Neonatal Nurs*. 1990;19:53-62.

Blackwell, R. and Chang, A., "Video Display Terminals and Pregnancy. A Review," *Br J Obstet Gynaecol*, 1988;95:446-453.

Bonde, J.P., "Semen Quality in Welders Before and After Three Weeks of Non-Exposure," *Br J Ind Med*, 1990;47:515-8.

Bonde, J.P., "Semen Quality and Sex Hormones Among Mild Steel and Stainless Steel Welders: A Cross Sectional Study," *Br J Ind Med*, 1990;47:508-14.

Bregmen, D.T., Anderson, K.E., Buffler, P., and Salg, J., "Surveillance for Work-Related Adverse Reproductive Outcomes," *Am J Public Health*. 1989; 79 Suppl:53-7.

Bureau of National Affairs, *Working Women's Health Concerns: A Gender at Risk?* (Washington DC: Bureau of National Affairs), 1989.

Bureau of Labor Statistics, *Back Injuries Associated with Lifting* (work injury report), Washington, DC, U.S. Government Printing Office, Bulletin 2144: August, 1982:1.

Bureau of Mines, Information about protective coverings for women and sizing charts for women's work clothes developed by the Texas Tech Textile Research Center can be obtained from: Jim Peay, Bureau of Mines, Cochroan Mill Road, P.O. Box 18070, Pittsburgh, PA 15236.

Canadian Pediatric Society, Infectious Diseases and Immunization Committee, "Cytomegalovirus Infection in Day-Care Centres: Risks to Pregnant Women," *Can Med Assoc J*. 1990; 142:547-9.

Cannon, L.J., Bernacki, E.J., and Walker, S.D., "Personal and Occupational Factors Associated with Carpal Tunnel Syndrome," *J Occup Med*, 1981, 23:255-258.

Chapman, J., Dame, L., and Rose, A., *A Citizen's Guide to Understanding Measurements of Toxic and Radioactive Contaminants*, March, 1990 Citizen's Environmental Coalition, 33 Central Ave., Albany, NY 12210.

Chavkin, W., ed., *Double Exposure: Women's Health Hazards on the Job and at Home*, NY: Monthly Review Press, 1984.

Chenier, N. Miller, *Reproductive Hazards at Work: Men, Women and the Fertility Gamble*, 1982, Canadian Advisory Council on the Status of Women, Ottawa, Ontario.

Clark, N., Cutter, T., and McGrane, J.A., *Ventilation: A Practical Guide*, 1984, New York, Center for Occupational Hazards.

Clarke, E., *Stopping Sexual Harassment: A Handbook*, Labor Education and Research Project, PO Box 20001, Detroit, MI 48220.

Coleman, M. and Beral, V. "A Review of Epidemiological Studies of the Health Effects of Living Near or Working with Electricity Generation and Transmission Equipment," *Int J Epidemiol*, 1988;17:1-13.

Cornell University, "MSDS at a Glance," pocket guide from Cornell University's Chemical Hazard Information Program.

Craig, M., *Office Workers' Survival Handbook: A Guide to Fighting Health Hazards in the Office*, London: BSSRS Publications, Ltd., 1981.

Daniell, W.E., Vaughan, T.L., and Millies, B.A., "Pregnancy Outcomes Among Female Flight Attendants," *Aviat Space Environ Med*. 1990;61:840-4.

Daniels, C., Paul, M., and Rosofsky, R., *Family, Work and Health*, Women's Health Unit, Dept. of Public Health, Boston, 1988.

Daniels, C., Paul, M., and Rosofsky, R., "Health, Equity, and Reproductive Risks in the Workplace," *J Public Health Policy*. 1990;4:449-62.

DeMatteo, B., *Terminal Shock–The Health Hazards of Video Display Terminals*, Toronto: NC Press Limited, 1985.

Dooley, P. and Tarlau, E., "Playing Industrial Hygiene to Win." Booklet available from P. Dooley, UAW S&H Dept, Detroit.

Epstein, S., *The Politics of Cancer*, NY: Anchor Press, 1979.

Florack, E. and Zielhuis, G. "Occupational Ethylene Oxide Exposure and Reproduction," *Int Arch Occup Environ Health*, 1990;62:273-7.

Food & Allied Service Trades, "Information Access: Health and Safety Documents You Have a Right to Under the Law," a Health and Safety Factsheet, AFL-CIO.

Food & Allied Service Trades (FAST), "Review of Women's Occupational Safety and Health Issues," AFL-CIO, 1988.

Forman, S.A. et al. "Psychological Symptoms and Intermittent Hypertension Following Acute Microwave Exposure," *J Occup Med*, 1982;24:932-934.

Forssman, S. and Coppee, G.H., *Occupational Health Problems of Young Workers*, Occupational Safety and Health Series, No. 26, International Labour Office, Geneva, 1984.

Fraser, T. Morris, *Ergonomic Principles in the Design of Hand Tools*, Occupational Safety and Health Series, No. 44, International Labour Office, Geneva, 1980.

FTQ, Service d'education, *Occupational Deafness Claims Guide*, FTQ, Service d'education, 2100 avenue Papineau, 4eme etage, Montreal Quebec Canada H2K 4J4 (available both in French and English).

Gordis, L., ed., *Epidemiology and Health Risk Assessment*, Oxford University Press, 1988.

Green, J., "Detecting the Hypersusceptible Worker: Genetics and Politics in Industrial Medicine," *Int J Health Serv*, 1983;13:247-264.

Guirguis S., Pelmear, F., Roy M., and Wong, L., "Health Effects Associated with Exposure to Anaesthetic Gases in Ontario Hospital Personnel," *Br J Ind Med.* 1990;47:490-7.

Hadler, S.C. et al., "Public Health Considerations of Infectious Diseases in Child Day Care Centers," *J Pediatrics*, 1984:105:683.

Hallenbeck, W.H. and Cunningham, K.M., *Quantitative Risk Assessment for Environmental and Occupational Health*, Lewis Publishers, 1986.

Hart, R.W., Terturro, A, and L. Nimeth, eds, "Report of the Consensus Workshop on Formaldehyde," *Environ Health Perspect*, 1984;58:323-381.

Helmkamp, T.E. et al., "Occupational Noise Exposure, Noise-Inducing Hearing Loss, and the Epidemiology of High Blood Pressure," *Am J Epidemiol*, 1985;121:501-14.

Huff, J.A., "Carcinogenicity of Select Organic Solvents," in *Proceedings of the International Conference on Organic Solvent Toxicity*, Stockholm, Oct. 1984.

IARC Monographs on the Evaluation of the Carcinogenic Risk of Chemicals to Humans. Chemicals, Industrial Processes and Industries Associated with Cancer in Humans, Supplement 4, International Agency for Research on Cancer, Lyon, 1982;7-264.

International Printing Pressman and Assistant's Union of North America, *Fighting Noise: A Manual for Worker Action*.

International Labour Office, *Education and Training Policies in Occupational Safety and Health and Ergonomics – International Symposium*, Occupational Safety and Health Series, No. 47, Geneva, 1982.

Joffe, M., "Male and Female Mediated Reproductive Effects of Occupations: The Use of Questionnaire Methods," *J Occup Med*, 1989;31:974-9.

Kazia, R. and Grossman, R.L., *Fear at Work: Job Blackmail, Labor and the Environment*, NY: Pilgrim Press, 1982.

Klitzman, S., Silverstein, B., Punnett, L., and Mock, A., "A Women's Occupational Health Agenda for the 1990s" *New Solutions*, 1990;1.

Knox, S. and Wallerstein, N. *Hazard Communication Manual: How to Implement Right to Know in Your Business*, Occupational Health Program of the New Mexico School of Medicine and New

Mexico Occupational Health and Safety Bureau, Albuquerque and Santa Fe, NM. 45 page looseleaf, single copies free.

Huel, G., Mergler, D., and Bowler, R., "Evidence for Adverse Reproductive Outcomes Among Women Microelectronic Assembly Workers," *Br J Ind Med*, 1990;47:400-4.

Kyyronen, P., Taskinen, H., Lindbohm, M.L., Hemminki, K. and Heinonen, O.F., "Spontaneous Abortions and Congenital Malformations Among Women Exposed to Tetrachloroethylenen in Dry Cleaning," *J Epidemiol Community Health*, 1989;43:346-51.

La Rosa, J.H., "Women, Work and Health: Employment as a Risk Factor for Coronary Heart Disease," 1988;158:6 Part 2:1597-1602.

Labor Institute, NYC. Source for informative Safety and Health posters.

LaDou, J., ed., "The Microelectronics Industry," *Occup Med: State of the Art Reviews*, 1986; 1:1 (entire issue).

Levy, B.S. and Wegman, D.H., (eds), *Occupational Health: Recognizing and Preventing Work Related Disease*, Boston: Little Brown, 1988.

Lewy, R., "Prevention Strategies in Hospital Occupational Medicine," *J Occup Med*, 1981;23:109-111.

Liebman S., *Do It At Your Desk: How to Feel Good from 9 to 5. An Office Workers Guide to Fitness and Health.* Tilden Press, Washington, DC, 1982.

Lindbohm, M.L., Taskinen, H., Sallmen, M., and Hemminki, K., "Spontaneous Abortion Among Women Exposed to Organic Solvents," *Am J Ind Med*, 1990;17:449-63.

Lockey, J.E. and Lemasters, G.K., *Reproduction, The New Frontier in Occupational and Environmental Health Research*. Liss, 1984.

Mackay, C., "Visual Display Units," *Practitioner*, 1989;233: 1496-8.

Marschall, D. and Gregory, J., *Office Automation: Jekyll or Hyde? Highlights of the International Conference on Office Work and New Technology*. Cleveland, Working Women Education Fund, 1983.

Manser, W.W. "Lead: A Review of the Recent Literature," *JPMA*, 1989;39:296-302.

MASSCOSH and the Boston Women's Health Book Collective, *Our Jobs, Our Health: A Women's Guide to Occupational Safety and Health*, 1983.

MASSCOSH, *Reproductive Hazards in the Workplace: A Resource Guide*.

McAteer, J. Davitt, *Textile Health and Safety Manual: A Complete Guide to Health and Safety Protection on the Job*, Occupational Health and Law Center, 1536 16th St., N.W., Washington, DC. 1986. $6.95 paperback.

McCann, M., *Artist Beware*, NY: Watson-Guptill Publications, 1979.

McManus, J., *The Deadly Dilemma: When OSHA Fails to Protect the Worker's Right to a Workplace Free of Health and Safety Hazards*, WISCOSH, 1334 S. 11 St, Milwaukee, WI 53204, 1987.

Merletti, F., Heseltine, E., Saracci, L. et al, "Target Organs for Carcinogenicity of Chemicals and Industrial Exposures in Humans: A Review of Results in the IARC Monographs on the Evaluation of the Carcinogenic Risk of Chemicals to Humans," *Cancer Res*, 1984;44:2244-2250.

Michel, S.E., "U.S. Supreme Court Agrees to Hear Fetal Protection Employment Policy," *Occup Health Saf*, 1990;59:40-1.

National Union of Hospital & Health Care Employees, "Watch Your Step: 6 Steps to A Safer Workplace," a Health Workers Guide.

National Union of Hospital and Health Care Employees, "Health Problems from Lifting and Carrying," Factsheet, (includes exercises for low back pain), RWDSU AFL-CIO, 330 W. 42 St., NY, NY 10036.

National Institute for Occupational Safety and Health (NIOSH), "Guidance for Indoor Air Quality Investigations," Hazards Evaluations and Technical Assistance Branch, NIOSH, January 1987.

Needleman, H.L., "What Can the Study of Lead Teach Us About Other Toxicants?," *Environ Health Perspect*, 1990;86:183-9.

Nelkin, D. and Brown, M. *Workers at Risk*, Chicago: University of Chicago Press, 1984.

Nelson, K.E. and Sullivan-Bolyai, J.Z., "Preventing Teratogenic Viral Infections in Hospital Employees: The Cases of Rubella, Cytomegalovirus and Varicella-Zoster Virus," *Occup Med: State of the Art Reviews*, 1987, 2:3:471-498.

Nelson, L., Kenen, R., and Klitzman, S., *Turning Things Around: A Women's Occupational and Environmental Health Resource Guide*, Washington, DC, National Women's Health Network, 1990.

Nine to Five Newsletter. Published 5x/yearly by 9 to 5, the National Association of Working Women. All members receive the newsletter. Yearly membership is $15-$25, based on your income. For nonmembers, subscriptions are: individuals - $25/ institutions-$40.

North Carolina Occupational Safety and Health Project and North Carolina Communications Workers of America, "Office Workers Stress Survey Results," March 1985.

Nurminen, T., "Shift Work, Fetal Development and Course of Pregnancy," *Scand J Work Environ Health*, 1989;15;395-403.

Nurminen, T., Jusa, S., Ilmarinen, J., and Kurppa, K., "Physical Work Load, Fetal Development and Course of Pregnancy," *Scand J Work Environ Health*, 1989;15:404-14.

NYCOSH, *Injured on the Job: A Handbook for NY Workers*.

NYCOSH, "Pregnant and Working: What are Your Rights?" 1985.

NYCOSH, "OSHA's Hazard Communication Standard," Factsheet, Spring 1988.

NYCOSH, *The VDT Book*, 1989.

Occupational Safety and Health Reporter. "Outlook: Occupational Safety and Health in 1989: A BNA Special Report," Vol. 18, #32 (Jan. 11, 1989).

Olsen J. and Hemminki, K. et al., "Low Birthweight, Congenital Malformations, and Spontaneous Abortions Among Dry-Cleaning Workers in Scandinavia," *Scand J Work Environ Health*, 1990;16:163-8.

Ontario Ministry of Labor, Monthly Library Bulletin on Occupational Health and Safety. Free of charge.

Participatory Research Group, *Short Circuit: Women in the Automated Office*, Toronto, Ontario, 1985.

Persaud, T., "The Pregnant Woman in the Workplace: Potential Embryopathic Risks," *Anat Anz*, 1990;170:295-300.

PHILAPOSH, "Workers' Rights vs. Industrial Interests: The Failure of OSHA in the 80's," a special report, July-August, 1983, #83 of *Safer Times*.

PHILAPOSH, "Ergonomics and Job Design," Factsheet, October 1988.

PHILAPOSH, "AIDS in the Workplace." Factsheet and conference materials. 1989.

PHILAPOSH, "Getting Management Records Under the OSHA Access Rule," Factsheet, March 1989.

PHILAPOSH, *Getting Job Hazards Out of the Bedroom: The Handbook on Workplace Hazards to Reproduction*. 1988.

Paul, M. and Himmelstein, J. "Reproductive Hazards in the Workplace: What the Practitioner Needs to Know About Chemical Exposures," *Obste and Gynecol*, 1988;71 (6 Part 1):921-928.

Pelmear, P., "Low Frequency Noise and Vibrations: Role of Government in Occupational Disease," *Semin Perinatol*, 1990;14: 322-8.

Retail, Wholesale & Department Store Union, "Information Access: Health and Safety Documents You Have A Right to Under the Law," a Safety and Health Factsheet, AFL-CIO.

Ratcliffe, J.M., Schrader, S.M., Clapp, D.E. et al., "Semen Quality in Workers Exposed to 2-Ethoxyethanol," *Br J Ind Med*, 1989;46:399-406.

Ricci, E., "Reproductive Hazards in the Workplace," *NAACOGS Clin Issu Perinat Womens Health Nurs*, 1990;1:226-39.

Rogan, Walter J., "Breastfeeding in the Workplace," *Occup Med*, 1986;411-413.

Rossol, M., *Stage Fright: Health and Safety in the Theatre. A Practical Guide*, Center for Occupational Hazards, 2 Beekman Street, NY, NY, 10038. $9.95 plus $2 postage, paperback.

Rowland, A., "Black Workers and Cancer," *Labor Occupational Health Program Monitor*, 1980;14.

Schrag, S.D. and Dixon, R., "Occupational Exposures Associated with Male Reproductive Dysfunction," *Annu Rev Pharmacol Toxicol*, 1985;25:567-592.

Selevan, S.B., Lindbohn, M.L. et al, "A Study of Occupational Exposure to Anti-Neoplastic Drugs and Fetal Loss in Nurses," *N Eng J Med*, 1985,313:1173-1177.

SEIU Health Safety Department, *The AIDS Book: Information for Workers* (mainly health care), 1313 L Street N.W., Washington, DC 20005 ($2.50 prepaid).

SEMCOSH, "How to Conduct a Workplace Health and Safety Inspection," Factsheet, 1988.

Savitz, D.A., John, E.M., and Kieckner, R.C., "Magnetic Field Exposure From Electric Appliances and Childhood Cancer," *Am J Epidemiol*, 1990;131:763-73.

Scott, A.J. and LaDou, J., "Shiftwork: Effects on Sleep and Health with Recommendations for Medical Surveillance and Screening," *Occup Med*, 1990;5:273-99.

Shortridge, L.A., "Advances in the Assessment of the Effect of Environmental and Occupational Toxins on Reproduction," *J Perinat Neonatal Nurs*, 1990;3:1-11.

Sorahan, T. and Waterhouse, J.A.H., "Cancer Incidence and Cancer Mortality in a Cohort of Semiconductor Workers," *Br J. Ind Med*, 1985;42:546-550.

Spake, A., "A New American Nightmare," *Ms*, March, 1986, pp. 35-82 (9 pages).

Special Report for Barbers and Cosmetologists, Health and Safety Program of the Food and Beverage Trades Department, AFL-CIO, 815 Sixteenth Street, N.W., Washington, DC 20006 (single copy is complimentary).

Stein, Z. and Hatch, M., eds., *Reproductive Problems in the Workplace, Occup Med: State of the Art Reviews*, 1986;3.

Stellman J. and Henifin, M.S., *Office Work Can Be Dangerous to Your Health (revised and updated editions)*, NY: Fawcett Crest, 1989.

Stellman, J., "The Working Environment of the Working Poor: An Analysis Based on Workers' Claims, Census Data and Known Risk Factors," *Women and Health*, 1987;12::83-101.

Stellman, J., *Women's Work, Women's Health: Myths and Realities*, NY: Pantheon, 1977.

Stellman, J. and Henifin, M.S., "No Fertile Women Need Apply: Employment Discrimination and Reproductive Hazards in the Workplace," in *Biological Woman – The Convenient Myth*, eds. Hubbard, R., Henifin, M.S., and Fried, B. Cambridge, MA: Schenkman, 1982.

Stromberg, A.H., Larwood, L., and Gutek, B.A., *Women and Work: An Annual Review, Vol. 2*, Beverly Hills, CA: Sage Publications, 1987.

Stucker, I., and Caillard, J. "Risk of Spontaneous Abortion Among Nurses Handling Antineoplastic Drugs," *Scand J Work Environ Health*, 1990;16:102-7.

Suess, M.J., ed., *Nonionizing Radiation Protection*, (Series No. 10) Copenhagen, World Health Organization, Regional Office for Europe, 1982.

Teichman, R.F., Fallon, L.F., Jr., and Brandt-Rauf, P.W., "Health Effects on Workers in the Pharmaceutical Industry: A Review," *J Sociol Occup Med*, 1988,38:55-57.

The National Committee for Clinical Laboratory Standards (NCCLS) "Protection of Laboratory Workers from Infectious Disease Transmitted by Blood and Tissue; Proposed Guideline," NCCLS Document M29-P, 1987.

Thomas, J.A. and Ballantyne, B., "Occupational Reproductive Risks: Sources, Surveillance, and Testing," *J Occup Med*, 1990;32:547-54.

Tobey, S., "Bring in Your Own Expert," *Safer Times* (PHILAPOSH newsletter), April 1988.

U.S. Department of Labor, *Noise Control: A Guide for Workers and Employers*.

U.S. Department of Health and Human Services, *Working in Hot Environments*, Washington, DC, 1986 DHHS pub. # (NIOSH) 76-203.

U.S. Department of Health and Human Services, *Shiftwork and Health*, Washington, DC 1986, DHHS Pub # (NIOSH) 76-203.

Walsh, J., "Hi-Tech Health Hazards," *In These Times*, Oct. 16-22, 1985;10-11.

U.S. Congress, Office of Technology Assessment, *Reproductive Health Hazards in the Workplace*, Washington, DC, 1985. U.S. Government Printing Office, OTA-BA-266.

UAW, *What Every UAW Representative Should Know About Health and Safety*, guidebook, 4th printing, 1988.

UAW Purchase and Supply Department, *The Case of the Workplace Killers: A Manual for Cancer Detectives On the Job*, $1 from 8000 E. Jefferson, Detroit, MI 48214.

UAW Purchase and Supply Department, *Strains and Sprains: A Worker's Guide to Job Design*, $2, 8000 E. Jefferson, Detroit, MI 48214.

United Auto Workers Safety and Health Department, "Evaluating Right-to-Know Training Programs," *United Auto Workers Safety and Health Newsletter*.

Upton, A.C., "The Biological Effects of Low Level Ionizing Radiation," *Scientific American*, 1982;246:41-49.

U.S. Department of Labor, "Access to Medical and Exposure Records," OSHA Document # 3110, 1988.

U.S. Department of Labor, Office of the Secretary, Women's Bureau, *Women and Office Automation: Issues for the Decade Ahead*, 1985.

U.S. Congress, Office of Technology Assessment, *Automation of America's Offices*, Washington, DC, 1985.

U.S. Department of Health, Education, and Welfare, 1977, National Institute for Occupational Safety and Health Research Report, *Guidelines on Pregnancy and Work*, Washington, DC.

U.S. Department of Labor, "Brief Guide to Recordkeeping Requirements for Occupational Injuries and Illnesses," OSHA, June 1986.

VDT News: The VDT Health and Safety Report. Published bimonthly, subscriptions are $87.00 for one year/$150.00 for 2 years.

Weaver, C., "Toxics and Male Infertility," *Public Citizen*, 1986;7.

WHO, "Visual Display Terminals and Workers' Health," World Health Organization, WHO Offset Publications, 1987:99:1-206.

Wilk, V.A., *The Occupational Health of Migrant and Seasonal Farmworkers in the United States*, Farmworker Justice Fund, 1986, $15. (Order from National Rural Health Care Association, 301 E. Armour Blvd., Suite 420, Kansas City, MO 64111.)

Wilkerson, M., "Working Women in the United States," in *Black Working Women*, Berkeley: Center for the Study, Education and Advancement of Women, University of California, 1982.

Williams, L.A., *Reproductive Health Hazards in the Workplace*, Science Information Resource Center, 1988.

Wolff, M., "Occupationally Derived Chemicals in Breast Milk," *Am J Ind*, 1983;4:259-281.

Women in the Workplace: Conference Proceedings. Volume One: Occupational Health Issues, 1986, MA Dept. of Public Health.

Zielhuis, B.L., *Health Risks to Female Workers in Occupational Exposure to Chemical Agents*, Springer-Verlag, 1984.

Ziem, G., "TLVs: How Good Are They? *Philaposh Factsheet*, Philaposh Project on Safety and Health, February 1988.

Index